Process Management
A Guide for the Design of Business Processes

Springer

Berlin
Heidelberg
New York
Hong Kong
London
Milan
Paris
Tokyo

Jörg Becker · Martin Kugeler
Michael Rosemann
Editors

Process
Management

A Guide for the Design
of Business Processes

With 83 Figures
and 34 Tables

 Springer

Prof. Dr. Jörg Becker (E-Mail: becker@wi.uni-muenster.de)

Westfälische Wilhelms-Universität Münster
Institut für Wirtschaftsinformatik
Leonardo-Campus 3
48149 Münster, Germany

Dr. Martin Kugeler (E-Mail: martin@kugeler.de)

Am Steintor 31a
48167 Münster, Germany

Prof. Dr. Michael Rosemann (E-Mail: m.rosemann@qut.edu.au)

Queensland University of Technology
School of Information Systems
2, George Street
Brisbane QLD 4001, Australia

Translated by
EDV-Studio Kortheuer GmbH (WWW: http://www.edvkort.de/)
Haus Gravener Straße 100
40764 Langenfeld, Germany

ISBN 3-540-43499-2   Springer-Verlag Berlin Heidelberg New York

Cataloging-in-Publication Data applied for

A catalog record for this book is available from the Library of Congress.

Bibliographic information published by Die Deutsche Bibliothek
Die Deutsche Bibliothek lists this publication in the Deutsche Nationalbibliografie;
detailed bibliographic data available in the internet at *http.//dnb.ddb.de*

Springer-Verlag Berlin Heidelberg New York
a member of Springer Science + Business Media GmbH

http://www.springer.de
© Springer-Verlag Berlin Heidelberg 2003
Printed in Germany

Cover design: Erich Kirchner, Heidelberg

SPIN 10992234   42/3111 – 5 4 3 2 1 – Printed on acid-free paper

Preface

Managing business processes is a necessity for every organisation. Since the early work of Adam Smith, this challenge caught the interest of organisations as well as that of the academic world. However, it wasn't until the fundamental contribution of Michael Hammer, that an entire new paradigm for process management was created. The proposed radical focus on business processes led to new organizational structures and IT-related solutions.

From an organizational viewpoint, new areas of responsibilities have been designed. The classical job enrichment and job enlargement have been applied along business processes. As a consequence, process owners, process managers and most recently even Chief Process Officers (CPO) have been institutionalized as an approach to appropriately acknowledge the requirements of process-oriented organization.

IT-related developments include the uptake of workflow management systems that automate the execution of business processes. Moreover, the worldwide focus on implementing Enterprise Systems was often motivated by the intention to provide an integrated application for the business processes of an organization. However, many of these IT-focused approaches towards process management failed because they were dominated by the complexity of the selected software solution rather than concentrating on the challenge of true business / IT alignment.

The academic progress in all these areas is well-advanced. Nevertheless, until today publications that translate the elaborated scientific state of the art into information that addresses the requirements of practitioners are still the exception. On the one side, most of the related literature simply states the perceived advantages of business process management without sufficiently describing ways for the operational implementation of this approach. On the other side, many challenges of practical process management projects, such as dealing with hundreds of process models or the efficient communication of project results, are still not comprehensively discussed from a more academic viewpoint.

It is the aim of this book to further fill this gap. Our aim is to present a comprehensive procedure model with high practical applicability. A main feature of this model is its focus on business process modeling. This book discusses the main project phases covering the goals of process management, project management, the design of company-specific modeling guidelines, as-is and to-be modeling, process-oriented re-design of the organizational hierarchy, process implementation and change management. A final chapter elaborates on specific application areas such as benchmarking, knowledge management or Enterprise Systems.

The overall design of this book is based on the guideline "As much practical relevance as possible, as much theory as required." A pivotal role is played by a consistent case study which can be found throughout the book and which rounds up the more academic discussions of each chapter.

Thus, we are very grateful to DeTe Immobilien, the organisation which was willing to share its journey towards a more process-oriented organisation with us. For more than four years we have been collaboratively working in this area. The support of this organisation is allowing us to report not only on the successes, but also on the constraints and errors of this project. Members of the underlying project were also willing to actively contribute as co-authors to the single chapters of this book. This was a main factor for the practical relevance of the ideas that are discussed in this book.

Editors of a book always carry a consolidating role. The actual success, however, depends on the contributions of the author. Consequently, we would like to express our appreciation to each author not only for the chapters, but for the careful integration in the overall context of this book. Mr Patrick Delfmann significantly supported this book project with his detailed work on the final design of the manuscript. We are especially grateful to the detailed work of Mrs Roswitha Kortheuer and Ms Islay Davies for their translation of this publication.

We hope this book will be inspirational for a wide audience and provides a convincing statement for the idea of business process modeling.

Münster / Brisbane, March 2003   Jörg Becker
                                                              Martin Kugeler
                                                           Michael Rosemann

Contents

The Process in Focus

Project Management

Preparation of Process Modeling
Michael Rosemann **41**

From Strategy to the Business Process Framework
Jörg Becker, Volker Meise **79**

As-is Modeling and Process Analysis
Ansgar Schwegmann, Michael Laske **107**

To-be Modeling and Process Optimization
Mario Speck, Norbert Schnetgöke...**135**

Design of a Process-oriented Organizational Structure
Martin Kugeler, Michael Vieting **165**

Process Implementation – Process Roll-out

Continuous Process Management

**Additional Application Areas and Further Perspectives –
Beyond Re-engineering**
Michael zur Mühlen..**251**

Figures

Tables

The Process in Focus

Jörg Becker, Dieter Kahn

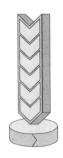

1.1
Environment

Changes in the economic environment force companies to constantly evaluate their competitive position in the market and to search for innovations and competitive advantages. Companies most often separate the external view of their company into a) the task environment, i.e. the direct relationship to their business partners; and b) to the global market environment where the activities of the [other] companies are only of little influence.[1]

Task environment and market environment

The task environment of a company is formed by the buying behavior of the customers, by the market structures, and by the dynamics of competition. An example of changed buying behavior is the increasing demand for individualization of products and services that leads to higher market fragmentation. The related increase of variants in production results in higher coordination costs for procurement, production, services, distribution and disposal.[2] An attempt to master the increasing complexity by applying additional coordination mechanisms may lead to the so-called complexity trap, when the overhead costs for additional control and coordination systems surpass the profit gained by the multitude of variants.[3]

Changes in the task environment

Complexity trap

In addition to the focus on the external view of a company, in terms of market, product line, quality of services and customer satisfaction, the focus on the internal view of a company increases as well, i.e. on the efficient and innovative execution of activities within a company. Stalk, Evans and Shulman emphasize the necessity to focus on the corporate dynamics:

From the external view to the internal view

---

[1] See Meffert (1998), p. 28.
[2] See Adam (1997), p. 25.
[3] See Adam, Rollberg (1995), p. 667; Adam, Johannwille (1998), p. 5.

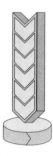

"At the time when the economy was still relatively static, the strategy could be static as well. In a world of long-life products, stable consumer needs, clearly delimited national and regional markets and recognizable competitors, competition was a "positional warfare" where companies took certain positions on a chess-board [...]

Today, competition is a "mobile warfare", where success depends on the anticipation of market trends and on quick response to changing customer needs. Successful competitors quickly develop products, markets and sometimes even complete branches, with a view to leaving them as quickly again – a procedure which resembles an interactive video game. In such an environment, the core of the strategy is to be found not in the structure of the products and markets of an enterprise, but in the dynamics of its behavior."[4]

Focus on corporate dynamics

1.2
From function orientation to process orientation

Optimization of individual functions

In recent decades, companies have orientated to an efficient execution of individual functions, which has led to a local optimization and perfection of functional areas. Technological and organizational developments, for example, have caused a significant increase in productivity and quality in the areas of accounting, logistics and production through the use of new information and communication technologies such as standardized software, call centers, Inter- and Intranets, as well as through the execution of organizational concepts such as the outsourcing of functional areas. Local optimization, however, caused the interrelationships of the operational functions to retreat into the background. Unfortunately, the costs for coordination between individual areas of the company increase with the autonomy of functional areas. The use of modern information and communication technologies does not eliminate this structural problem. A company's internal electronic communication just reduces the symptoms, i.e. the duration of the coordination processes. The strengthening of a company in its totality and the reduction of existing interfaces, however, require a focus on the cross-functional business processes. The idea of a process-oriented corporate design is not new at all. It has been given increasing attention through the catchwords "Business Process Reengineering" and "Business Process Management" since the end of the 80's. Even from as far back as the early 30's, Nordsieck

Effects of information and communication technology

---

[4]   Stalk, Evans, Shulman (1992), p. 62.

pointed to the necessity of a process-oriented corporate design and continued to do so in 1972:

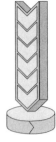

Process orientation: early approaches...

"[The division of corporate tasks] requires, in any case, aiming at a clear division of the processes. This division has to be in line with the targets, the development of the [process-] objects and, in particular, with the rhythm of the tasks."[5]

"[...] The operation [is] really a permanent process, an uninterrupted chain of performances [...]. The real structure of the operation is that of a river. It continuously creates and distributes new products and services based on the same tasks or on tasks which only vary in details. [...]

The question in view of this overall thinking is how to divide the tasks of a company other than by natural technical process phases?"[6]

In spite of the early discussion of this topic in the academic literature, process orientation only started to be practiced by companies in the 80's, after GAITANIDES[7], SCHEER[8], PORTER[9], DAVENPORT[10] as well as HAMMER and CHAMPY[11] had published their approaches.[12]

...late popularity

This book differs in its objective from the above-mentioned literature in a number of areas. On the one hand, a number of concepts already exist about process organization. They deal with the theoretic fundamentals of this organizational form, but give little guidance on how to convert the concepts into corporate practice. The following reference to HAMMER, CHAMPY and the practical conversion of their concept of Business Reengineering illustrates this observation: "Therefore, we have not entered into details on how companies should convert Business Reengineering in practice."[13] On the other hand, the works of (primarily) American authors are characterized by numerous practical examples, which, however, do not form part of an overall concept.

Objective of this book

This book tries to close this gap by presenting the general, conceptual principles for the introduction of a higher process orientation in the form of a process model. Furthermore, this book is based on a complete case study which facilitates the practical handling and conversion into practice.

Design of executable concepts

---

[5] Nordsieck (1934), p. 77.
[6] Nordsieck (1972), column 9.
[7] See Gaitanides (1983).
[8] See Scheer (1990).
[9] See Porter (1989).
[10] See Davenport (1993).
[11] See Hammer, Champy (1993) and Hammer (1996).
[12] Körmeier gives an overview of the history of process-oriented corporate design (1995).
[13] Hammer, Champy (1993), p. 216.

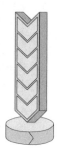

1.3
The term of process

Subject of process orientation

The business processes of a company are the central point of the process-oriented corporate design. While the organizational structure divides the company into partial systems (e.g. departments, divisions, units) with their assigned tasks[14], a business process deals with the execution of these tasks as well as with the coordination of their timely and other aspects (who does what, how and with what).[15] Elementary components of a task are those activities that form the basic parts of a (working) process. An activity and / or a function is a working step which has to be executed in order to render a service.

Process

A process is a completely closed, timely and logical sequence of activities which are required to work on a process-oriented business object.[16] Such a process-oriented object can be, for example, an invoice, a purchase order or a specimen. A business process

Business process

is a special process that is directed by the business objectives of a company and by the business environment.[17] Essential features of a business process are interfaces to the business partners of the company (e.g. customers, suppliers). Examples of business processes are the order processing in a factory, the routing business of a retailer or the credit assignment of a bank.

Porter's value chain

PORTER presented his model of a value chain in 1980 and contrasted corporate activities by primary activities and by supporting activities.[18] Primary activities are value-creating activities with a direct relation to the manufactured product; they therefore contribute to the economic outcome of the company, i.e. activities in the area of procurement, operations (i.e. production), marketing and sales, logistics and customer services. Supporting services do not have a direct relation to manufactured products and services; however, without these supporting services no value-creating activities could be executed. Examples of supporting areas are human resource management, accounting and data processing. Consequently, a core process is a process whose activities directly relate to the product of a company and therefore contribute to the creation of value in a company. In contrast to the core process, a sup-

---

[14] See Lehmann (1974), column. 290.

[15] See Schweitzer (1974), column. 1; Esswein, (1993), p. 551.

[16] See Becker, Schütte (1996), pp. 53; Rosemann (1996a), p. 9.

[17] See Nordsieck (1972), pp. 8-9, who defines the business process (the total of all business processes of a company) as a stepwise realization of the corporate goals.

[18] See Porter (1989), p. 63. Another work of Porter is discussed in chapter 4.3.3.

porting process is a process whose activities do not create value from the customer's point of view, but are necessary in order to execute a core process. The boundaries between core- and supporting processes are floating since the same process can be a core process or a supporting process, depending on different contexts and different enterprises. The term "supporting process" must by no means be regarded as carrying little importance. On the contrary, supporting processes are essential requirements to execute the core processes; they merely do not have any direct contact points to the processed products and / or services. Without supporting processes, the execution of the core processes would be impossible. For this reason, they are also called enabling processes. In addition, supporting processes can migrate to core processes. Retailing companies, for example, do not execute logistic tasks in the core process of central regulation but concentrate on regulation activities that represent supporting processes in a typical warehouse business.

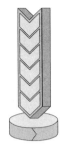

Core process and supporting processes

1.4
Consequences for management – six theses for process orientation

1.4.1
Every organizational business structure has its own efficiency relation

The target of every economically-working company is to make a profit. The manner, form and scope in which this target can be reached are determined by the structures and processes of the rendered services. The value-creating process is simultaneously subject to the laws of the market – a company can only be successful if the customers accept the services of this company and not those of the competition, and if the customers pay for these services. The permanent environmental influences on the business processes, however, often lead to self-dynamics that move the original targets of the business activities to another level.

Changing influences

Scope and complexity of the corporate structures, as well as the type of management and cooperation, determine the internal and external image of a company in different ways. A company is a living social and technical system and is therefore permanently impacted by influences from different directions, which are reflected in the development process of the company. Social factors such as psychology, biology, emotion and despotic thinking are factors in the corporate organization that must not be underesti-

Emotional intelligence vs. analytic intelligence

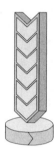

mated. Indeed, management often ranks emotional intelligence higher than analytic intelligence.

1.4.2
The survival pressure in competition causes sensible adaptation processes to develop

Competition guarantees evolution

Competition guarantees the evolution of companies and markets since there will always be competitors with new ideas or innovations. In profitable business fields, competition is adequately strong.

Process models guarantee adaptability

Organizations who want to remain competitive need suitable guiding, controlling and auditing instruments, as well as a transparent model of their own value chain, which can be adapted to the constantly changing conditions at anytime. This requires them to be fast and cost effective. Process benchmarking is one possible way to learn from competition. By identifying exemplary business processes adopted by competing organizations a company is able to gain an important basis for a profitable process design. This requires, however, that the company is well organized. An effective comparison with the competition is only possible if the internal structures are such that comparison and measurement can be made.

Benchmarking

This will not usually be valid for the business process in its entirety, but more for its specific partial processes. A comparative market orientation, for example in logistics, is efficiently performed with logistic companies, as they are specialists in this business line. The comparison here is not restricted to the original business line of the company, but directs toward the organization that has the greatest experience potential (best practice). Surprisingly, the market participants and the competitors in turn assist the realization of their own corporate goals.

1.4.3
People and their environment determine the process changes

Integration in favor of the customer process

An important aspect of the process-focused way of thinking is the horizontal cooperation of all participants, i.e. working in concert with the customer process. The most successful companies often have the highest degree of integration of people involved in the process.

Networks instead of monolithic companies

Advanced corporate strategies assume that future companies will be structured in "networks", and they visualize virtual enterprises with scattered, decentralized single activities. These enterprises are more flexible and more capable of learning than traditional companies. They enable a creative participation of many employees inside and outside the company, and motivate and generate more innovations and better performances.

The current complex world has to be managed in a way that differs from the management of yesterday. Networked organizations and business processes covering multiple locations require a "networked management". Processes have to be controlled by overlapping workgroups, which requires related feedback between the groups. The responsibility for the process has to be part of the process itself. Those responsible for a function within a process closely cooperate with the process owners, i.e. those responsible for the whole process. Both parties therefore take over a common responsibility. The individual working steps are additionally linked by an interdisciplinary cooperation that allows restriction of external control and consequently reduces indirect costs.

Networked management through decentralized coordination

Process management requires a new and better form of corporate management. The orientation exclusively directs toward the customer rather than toward the supervisor. It is the "customer process" that is binding. The efficiency of the process is measured by the customer himself, and not by controllers from within the company. People in the process are motivated by additional responsibility, autonomy in their activities, and more experiences of success. However, all those involved in the process have to be informed of the development and its target. Therefore, it becomes more and more important to transfer visions, strategic guidelines, and operative goals to all employees via suitable communication and training. The corporate culture and philosophy is to initiate an integral destination finding process in which all participating units in the company have to be involved. Any fear of new things causes a hindrance and has to be eliminated. "Hinderers" can block a lot of power; therefore any blockages have to be found and removed. Furthermore, Change Management has to be practiced at all corporate levels – from top management down to the lowest operative level in the company's hierarchy.

The customer is the decisive benchmark

Motivation to cooperation

1.4.4
Flexibility guarantees a top position

The current society does not only request top performances but also has to be permanently capable of adapting to new conditions. Only companies with high flexibility will survive in the long-term. This requires knowing what has to be adapted and to what. Unstructured processes can hardly be adapted reasonably because control of the "side effects" is lost. Clearly structured processes show their interdependencies with other processes and can therefore be changed in partial sections without losing the overall context.

Identification of interdependencies

Companies often focus on their core competencies and favor out-sourcing parts of their own value chain to cooperative partners. This, however, requires the highest degree of integration and coordination. These tasks also have to be mastered, in principal, in well-structured process chains. Furthermore, responsibility for processes can virtually span across to the borders of the whole company. As such, there is a need to integrate the databases that contribute to the mastering of the processes; in particular with the process interfaces.

1.4.5
High Innovation potential and its effective use lead to top positions

Problems of traditional proposal systems

Proposal systems and / or the management of ideas are already established functions in numerous companies. The experiences of managers, however, have shown that it can be very difficult to activate this innovation potential among employees using these mechanisms. Time-consuming decision processes and resistance of the responsible persons block valuable innovation potential. In addition to the desire of employees to maintain possession of a function or process, it is the lack of knowledge in the cause-effect relationships within the company that is one of the major reasons for this attitude. The operational targets, the interdependencies of the processes, and the related process- or cost relevant effects are not recognized and – even worse – their identification is not sufficiently trained.

Knowledge of processes enables innovation potentials

Structured processes allow detailing down to the functional level and make relationships between the individual working steps transparent. This enables employees to understand the process, to make sense of it and – even more importantly – to improve their area of responsibility with good ideas. This way, the innovation potential within the whole company can be unleashed.

1.4.6
The ability to integrate all participants in the process secures the success

Changing the organization, IT and management tasks

As the requirements of a process-focused organization differ to those of a traditional enterprise that is structured by functions, the process orientation has certain consequences for management. The apparent changes in orientation relate to the organization of the company, the technical infrastructure, and the management tasks.

In regard to the organization of a company, the process orientation tends towards increasing relocation of competencies to lower hierarchy levels and, consequently, with larger "freedom of decision", but also with larger "areas of responsibility" of the individual employees. Through the combination of functionally-separated – but process-related – tasks, the employees get an inside view of the areas preceding and following their own activity. This conversion in the nature and scope of an employee's role should be supported by management through the creation of guiding principles as well as by stimulation of the individual employee to take over more responsibility. Furthermore, transfer mechanisms are needed to change the existing functional business structures. This requires an organizational feeling of partnership and togetherness, as well as the capability to succeed and the power to carry things out.

Guidelines for changes

However, the process orientation is not only a challenge for management in an organizational respect, but technological changes have to be handled as well. The existing information systems have to be evaluated to see whether or not they are suitable to support the processes, and whether or not they have to be adapted or even replaced. The process orientation is also a prerequisite for the use of technological potentials, which, in a function-oriented company, are difficult to convert or cannot be converted at all. This means that technologies such as workflow management systems, for example, can only be profitably used when their introduction is accompanied by organizational actions.

Information - technological changes

The demand for flexibility in the sense of a fast adaptation to the changing market structures and customer requirements, the demand for extended product lines, as well as the demand for efficient services in the company, is often the catalyst for reorganization projects. The measurement of actually-realized improvement potentials by traditional controlling instruments is in most cases only possible with restrictions. Therefore, new methods to evaluate a company have to be applied in order to judge the progress and to calculate the already achieved improvements.

Target: Flexibility

Measurement of improvement potentials

1.5
The case study – DeTe Immobilien

The following chapter explains a procedure for the process-oriented corporate design based on a case study in a facility management enterprise, the Deutsche Telekom Immobilien und Service GmbH. This case study resulted from the scope of a 2-year reengineering project which has been executed by DeTe Immobilien in cooperation with the Institut für Wirtschaftsinformatik of the Westfälische Wilhelms-Universität Münster, Germany. This project is one of the largest modeling and reorganization projects that has been executed in the Federal Republic of Germany up to now. 13 fulltime method experts, as well as more than 70 special

2-year reengineering project

Business-wide process modeling

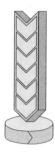

Processes, functions, technology

line experts, have collected, analyzed, optimized, consolidated and incorporated all business processes of the company DeTe Immobilien over a period of two years and have taken the new process versions into operation. In addition to pure business process modeling, partial projects have been executed in individual functional areas of the company, such as the conceptual design of a service management system, the introduction of SAP R/3 or the preparation for certification pursuant to ISO 9000ff and 14000ff. The analysis included not only company-internal processes, but also investigated in particular interfaces to the key accounts of DeTe Immobilien, i.e. the Deutschen Telekom AG, where the project team modeled a group-internal process which provided the facility infrastructure. The experiences gained within the scope of this project have been included in this book and the questions and concepts have been answered and / or explained and tested for feasibility in practical application.

1.5.1
Corporate culture

DeTe Immobilien: Facility management, branch of Deutsche Telekom AG

Merge of DeTeBau and ISM

DeTe Immobilien GmbH with headquarters in Münster / Westfalia, Germany, is the largest branch of Deutsche Telekom AG, with approx. 10,500 employees. DeTe Immobilien provides facility management services, i.e. development, planning, construction and management of facilities. In 1995, different areas of Deutsche Telekom AG, which were directly and indirectly involved in construction and management of the Telekom facilities, were integrated into the division dealing with facilities and service management, and "Immobilien und Service Management" (ISM). At the same time, the existing branch, DeTeBau, which was primarily involved with the maintenance of new constructions and / or reconstructions of the Telekom facilities in the new states in East Germany, became the DeTe Immobilien GmbH. A short time thereafter, the ISM division was integrated into the new company. DeTe Immobilien has an annual revenue of more than four billion Euro, the major share of which comes from the company's key account, Deutsche Telekom AG. In order to prepare the company for the free market, the project "PROFIT" started in 1996, which had the task, among others, of arranging for the process management and for the reconstruction of DeTe Immobilien. The scope of this project included the project "Process Modeling", the results of which have been reported in this book.

Organizational structure

The organizational structure of DeTe Immobilien in 1996 included 12 subsidiaries, and covered the Federal Republic of Germany with 108 service logistics centers and 350 external offices. The structure is designed primarily to guarantee a start time of 45 minutes for nearly all tasks even outside the normal office hours. The wide coverage and / or presence in the Federal Republic is a considerable competitive advantage for the company.

The processes in the individual subsidiaries are executed in the same way since the subsidiaries show almost identical organizational structures, and are able to provide all products from DeTe Immobilien. DeTe Immobilien manages a total of 34,000 facilities, from modern office buildings via antenna carriers ("radio towers") up to strategic computer centers, and service approx. 200,000 facility users.

Same processes in all subsidiaries

1.5.2
Products

Within the scope of the Telekom-Intranet, DeTe Immobilien GmbH — as of the end of 1996 — operates the largest Infranet in Germany, which is used for company-internal communication based on the Intranet and for the supervision of their facilities and infrastructure. DeTe Immobilien's services include the operation and optimization of facilities and technical equipment, as well as maintaining the stability of their value. In addition, the product line includes project processing, planning and the construction of buildings, as well as portfolio- and / or asset management on behalf of the customer. DeTe Immobilien distinguishes between an owner view and a user view. Relevant services for owners are, for example, the administration of rental contracts or energy management, while services for the users of the buildings consist of certain floor services such as the management of moving or cleaning services.

Infranet

Services of DeTe Immobilien

The combination of the latest building information and communication systems enables fast and efficient service processes to be provided. For example, the tasks are bundled in such a way that minimum driving, setup and working times result. The implementation of building information systems in a facility is an efficient means of executing error recoveries and operational tasks via remote control. In addition, error recoveries and preventive measures are combined, forming the basis for an efficient maintenance of value management.

1.6
Objective and structure of this book

This book will help to systemize the transition from a function-oriented to a process-oriented company and to support the conversion of such a project to operation. The book is divided into the

Target of this book

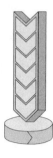

phases of the executed reorganization project and represents a guideline for the verification of already-executed works, and for the planning of pending tasks based on the individual project phases.

Contents of this book

This introductory chapter is followed by the second chapter, which explains the principles of project management. The third chapter describes the methodic basics for process modeling. In addition, examples of model guidelines and conventions are given in the appendix. The fourth chapter contains considerations of possible strategic alternatives and describes a system framework that can be taken as a guideline for the process modeling to follow. Chapters 5 to 9 describe the individual phases of the reorganization project: As-is modeling and analysis, to-be modeling and process improvement, design of the organizational structure, and

Contents of chapters

introduction of the processes, as well as the establishment of a continuous process management. Finally, the 10th chapter points to additional, innovative application areas for the results that come from a process modeling project. These areas can then either be understood as a subsequent project following a reorganization project, or as immediate and independent motivation for the execution of a process modeling project.

The essential conceptual principles are explained in all of the following chapters, as well as the working steps that are required for conversions, and the changes that result from them. Practical examples from the modeling project of DeTe Immobilien in the form of case studies show the related executions. Every chapter closes with a checklist for practical use in process management projects.

Project Management

Jörg Becker, Dieter Kahn, Clemens Wernsmann

Detailed reflections about how to manage any planned project are required to execute a project successfully. This is especially valid for modeling projects, since here, the classical tasks of project coordination are supplemented by defining and establishing methodologies. In general, projects represent tasks which are timely, limited, complex and interdisciplinary.[19] A successful execution of projects requires individual partial tasks and the utilization of personnel and resources to be organized, planned, controlled and verified. These functions are tasks of the Project Management.[20]

2.1
Project goals

Goals are essential for planning, controlling and auditing of projects. These goals have to be made clear to all project members in order to support the members in their target-oriented decisions.

The project goal includes the targets, i.e. the purpose of the project, such as product development, implementation of an IT-infrastructure, or the creation of a corporate process model. Next, the formal parameters have to be defined in terms of costs and time schedules.

Systemization of project goals

While the formal parameters are identical in most cases (cost minimization, observance of project runtime), the targets vary considerably. Process modeling processes serve superordinated projects to implement and / or to further develop the process management. At the top level, targets can be distinguished between organizational and IT-related goals.[21] Typical applications of process models within an organization management are the documentation of existing and projected processes, the benchmarking of

Performance goals

---

[19] See Schulte-Zurhausen (1995), p. 343.
[20] See Krüger (1994a), p. 374.
[21] A detailed discussion of the different purposes of process modeling is given in chapter 3.

Modeling purposes

processes and the certification pursuant in accordance with to DIN ISO 9000ff. The design of an information system is to document the system requirements (specifications), to customize integrated Enterprise Systems, as well as to specify workflows and to create simulation models.

Considering the magnitude of possible modeling purposes, it becomes clear that one single model will not be able to cope with all these purposes. Therefore, models to be taken as a basis for workflow management systems are more detailed than models for organizational descriptions. The first require among others roles and data structures, while the latter consider other areas (staff members, process costs).[22] Therefore, the purposes of the modeling project must be made clear from the very beginning.

Time goals

Based on the project goals, dates and costs must be planned and determined as standard values, or "milestones". The time schedule must include the desired end date of the project. A detailed plan has to include individual standard dates, so-called "milestones", which are calculated based on individual activities in the project plan and on available resources. Through detailed planning of individual milestones, the end date of the project can be verified and / or falsified. When planning the project, the most important item is the integrated consideration of dates and resources, since otherwise, no realistic standards for these milestones and for the end of project can be assigned. A special problem with the modeling project is the in-time availability of experts. One possibility is to free these experts entirely from their operative tasks and to assign them exclusively to the modeling project. Unfortunately, however, in most cases these experts are indispensable for their operative tasks. The difficulty in balancing the availability of experts between operative tasks and project activities has to be considered in the project plan, and uncertain factors in the time schedule have to be compensated for by time buffers. Especially, in case of conflicts between the time periods, the project and the project participants need a directive from management to know whether preference has to be given to the project or to the operational tasks.

Cost goals

Generally, the planning of costs is a very uncertain procedure. The types of costs in a project have to be identified and planned individually. In particular, when planning for personnel costs, sufficient human resources must be available in order to meet the time schedule. Moreover, speeding up completion of the project has to be compared with the added costs of additional personnel.

---

[22] See Chapter 6.2.1.

The initial situation of DeTe Immobilien prior to project start was as follows:

Due to the different business areas within DeTe Immobilien, the processes differed accordingly, and because there was lack of uniformity with the processes, the company suffered from losses of efficiency. These efficiency losses were even greater since the interfaces between the departments and the processes were not clearly defined. The process descriptions in the different departments of DeTe Immobilien were not uniform, the operating specifications showed gaps and different degrees of detailing, and there were even errors because of different documentation guidelines. The process orientation of the existing information systems was insufficient and, in addition, it was not embedded in an integrated architecture. Finally, the responsibility was delegated as function-oriented, and not outcome-oriented.

Situation of DeTe Immobilien prior to project start

Motivated by the recognized shortcomings, the project "Process Modeling" was initiated with primary focus on organizational objectives. Company-wide processes had to be created, i.e. consistent and integrated processes, from marketing via acquisition, quotations, contract conclusions, services, up to invoicing and general ledger.

Project goals

Assignments from the organizational population to the processes should be based on the "logics of the process". Friction losses resulting from organization and interfaces should be avoided, if possible. Ideally, the largest possible number of sequential working steps should be summarized in one organizational unit, as long as other reasons did not conflict, i.e. different competency requirements of succeeding worksteps or the loss of learning effects in case of a too large extension of tasks. This included the goal to improve the motivation of employees by process-oriented views and by allowing them to take over responsibility within the project. Every employee should, therefore, be able to recognize his share of contribution to the overall performance of the company and last but not least in the company's success.

The designation of process owners (i.e. designating positions with responsibility for certain processes) and the definition of clearly delineated process parameters should support the total management of the company.

The project goals are summarized as follows:

- Giving all employees a process-oriented view of all services of the company,
- Improvement of the business processes in terms of efficiency, redundancy avoidance, reduction of the processing time, and finally, decrease of process costs,

- Documentation of (improved) processes using business models, and improvement of communication between different organizational units,
- Clear definition of competencies,
- High motivation of employees,
- Increased capability of the company to respond to the market (flexibility).

Extended goals

The organizational changes to enable more of a process orientation should form the basis to develop process-oriented information systems. The created process models should support future insourcing and outsourcing decisions. Extended goals were processed which could be taken as the basis for workflow management systems (WFMS) and to be used as source information for personnel requirement planning. Another application of process models is to support the certification of the company (e.g. pursuant to DIN ISO 9000ff.).

2.2
Project plan

The project manager has the task of creating a project plan that includes the tasks to be executed, the planned durations of these tasks and their termination dates. In addition, he has to determine the approximate need for resources. Strategic parameters ("the new organizational structure has to be introduced on the 1st October next year") have to be synchronized with the time schedule of the project. If required, the project team has to be extended in order to be able to keep the defined deadlines.

Preparation of modeling

Typically, a modeling project includes phases – which is of course true for nearly every project. The modeling subject ("what" to model, e.g. the total enterprise or just a partial area, such as logistics), the perspectives ("for which purpose" to model, e.g. certification, software selection, organizational re-design) and the modeling methods and tools ("how" to model) are determined in the prestudy phase. In this phase, a glossary in the form of a technical model is designed which will then be completed continuously in all project phases to follow. In addition, it has to be determined which degree of detailing is adequate to the modeling purpose.

Strategy and business process framework

The starting point of the top-down approach in process modeling is derived from corporate strategy. At this point, a business process framework is created that includes the major functions of the enterprise on the highest level. This business process framework will then form the basis that supports navigation through the models and their retrieval.

Within the scope of as-is modeling, the states of the actual processes are collected and modeled as they currently exist. The as-is modeling serves not only to record the actual situation, but also to make the project team and the members of the involved business departments which belong (or will belong) to the extended project team, familiar with the modeling methods and modeling tools. The actual analysis will reveal shortcomings and enable a description of potential improvements.

As-is modeling

To-be modeling exploits the potential for process improvement that has been recognized in the actual analysis. New processes will be developed and modeled. If applicable, multiple steps have to be executed in order to migrate from the actual state to the to-be state, or it will explicitly be distinguished between the to-be model (what is possible under restrictions which cannot be eliminated within a short time) and the ideal model (what is theoretically the best model which, however, can only be converted on a mid-term or long-term basis).

To-be modeling

An important step in the process-oriented organization is deriving the organizational structure from the to-be process models. A logical process-oriented restructuring also includes a redesign of the organizational structure. This phase carries a certain volatile effect since in this phase the tasks are assigned to the organizational units, which confers power and influence to the assignment.

Organizational structure

The implementation phase converts the theoretic process improvements to real processes. This means that in the case of an organizational project a change in the processes often leads to a change in the organization. In the case of a process automation project the implementation of a workflow management system is required. In the case of a software development project programming and introduction of the software (as far as it is developed individually) and / or an adaptation (customizing) of the software and its implementation (as far as an Enterprise System is used).

Implementation

Even if the main restructuring project is terminated, the consideration of the process still has to remain the focal point. A process improvement in a continuous process management has to be understood as a process as well, which becomes an integral task of the operative management and which has to backup the competitive position for a long time.

Continuous process management

The procedure of a process-oriented organizational design is summarized in Fig. 2.1.

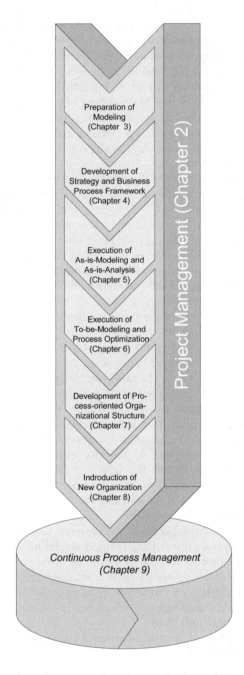

Fig. 2.1. Procedure of a process-oriented reorganization project

2.3
Project organization

As is the case in any other project, the process modeling project needs its own temporary organizational form to intersect with the "primary" organization of the enterprise. Skills (knowledge of the operational services, the required tasks, the actually valid process and possible improvements) and the methodic know-how (knowledge of modeling conventions and their application in the projects) have to be united in such a project. Skills are found in the departments, and methodic know-how in the organization department. Furthermore, external consulting companies or research institutes can be called upon to assist in the application of methods and in the execution of projects.

Features of project organization

The management appoints a project manager. It is advantageous to select a person from the organization department since this would eliminate a possible conflict if the project manager were said to be representing his own interests in the project. The role of the project manager in a process modeling project is extremely versatile.

Project manager

The results of the project will usually have a direct influence on the future processes of the project participants, especially if a reorganization project is concerned. In this respect, the project manager must have a certain feeling for the people involved. If he wants to avoid a project standstill – where everything remains as is – he must not fear any resistance from part of the individual groups – nor should he have to make decisions against the desires of certain groups. In order to counteract a defensive attitude by these groups, which may even result in a complete boycott of new procedures, he has to do a great deal of explanatory work to convince these groups of the benefits of the project and to gain their acceptance. The involvement of these groups in the project work is absolutely necessary. The project manager has to act in favor of integration instead of polarization, and he has to hold a singular view.

Tasks of the project manager

The project manager reports to the project steering committee that uses these reports to make decisions. The project steering committee (PSC) is composed of members of the board of directors and, within reason, of employees who will later be the project owners. Of similar importance is the early involvement of representatives of the staff committee in the PSC in order to observe their interests. The PSC holds periodic meetings, verifies the progress of the project based on the defined milestones, decides on the "stop or go" alternatives and makes the project-relevant and subject-related decisions.

Project steering committee

Project team

The project manager selects the individuals for the project team. On the one hand, this project team should be as large as possible in order to involve as many members of the company as possible. On the other hand, the project team should be as small as possible in order to create an effective working environment and to avoid the project team becoming a "debating circle". Finding a balance between "too large" and "too small" and creating the right mixture of special experts and method experts, are the most important tasks of the project manager when putting together the project team. In very large enterprises it seems to be reasonable to create a relatively small core team which can then be completely dedicated to the process. It has proven to be advantageous to nominate the process owners at an early stage, who can then act as specialized promoters of projected changes. They take over the role of thinking and acting ahead and are responsible for the enforcement of the process changes. In enterprises with multiple structures of equal type (e. g. multiple similar subsidiaries), they act as responsible multiplicators of new processes.[23]

Project organization of DeTe Immobilien...

... during as-is modeling

The project organization of DeTe Immobilien has not been constant over the total project time but has changed in the individual project phases. The only constant fact was the project steering committee (PSC) which was composed of members of management, members of the staff committee who accompanied the project over the total runtime with critical and constructive comments, and the project manager who came from the organization department.

In the actual analysis, three modeling teams were created who collected the procedures according to the existing organizational structure. These processes did not always match the processes of the business process framework[24] since the existing structures of the processes and organization have not been in conformity with this new framework. The modeling teams consisted of method experts, who asked process-related questions to the members of the departments in semi-structured interviews. The three modeling teams were assisted by a consolidation team who determined the modeling conventions required to support the consistency of the models. One manager from each department was named as steering coordinator. Consequently, there was one coordinator for the divisions "Facility Management", "Planning and Construction", "Portfolio Management and Sales" as well as "HR and Social Affairs" and "Administration and Finance".

---

[23] For additional tasks of the process owner refer to chapter 9.4.1.
[24] Refer to chapter 1 4.

The as-is modeling of the project organization was accompanied by the division "Strategy and Central Quality Management" (refer to Fig. 2.2).

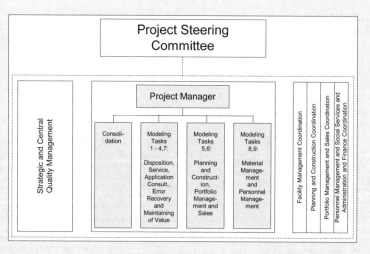

Fig. 2.2. Project organization of as-is modeling

The dual division of the as-is modeling organization, as already indicated, has been manifested in the organization of the to-be modeling phase. The project manager coordinated the method modeling experts. Two method experts were assigned to one user team in order to cope with the huge amount of tasks within a short time, and to guarantee the continuity of the modeling activities in case one of the method experts was not available.

... during to-be modeling

Every process-specific user team was headed by a process godfather who had the authority for decisions. This process godfather was nominated by the management and selected from the branch managers. The process godfather had to fulfill the following tasks:

- Naming of competent specialists for the interdisciplinary teams from different branches and from the head office.
- Guaranteed technical conversion of strategic directives.
- Technical quality assurance and decisions on process design.
- Securing of deadlines to be kept.
- Participation in the consolidation meetings of all process godfathers and execution of approvals in order to guarantee technical correctness.
- Agreement on alternative solutions and processes (including interfaces) with other process godfathers.
- Regulation of conflicts.

Tasks of the process godfather

Tasks of technical experts

The tasks of the technical experts nominated by the process godfather can be described as follows:

- Participation in workshops and interviews.
- Creation of team-specific approaches.
- Design of alternative solutions for to-be processes.
- Description of to-be processes on levels 2-n.
- Definition and / or editing of technical terms and creation of technical-term models.
- Execution of technical quality assurance.
- Agreement with other modeling teams on alternative solutions and processes, including interfaces.

Kick-off

When starting the modeling activities (so-called "kick-off"), each team held a 2-day workshop with all user experts, method experts, and the process godfather, in order to find a consensus with the targets, the methods and the procedures.

The kick-off events included individual actions as follows:

- To emphasize the importance of process orientation by a member of the management.
- To question the project members about their ideas and expectations, and to record their answers.
- To present the project "Process Modeling" together with the related goals and approaches.
- To present the recorded as-is processes to the modeling team.
- To explain and discuss the to-be processes of levels zero and one.
- To define additional team-specific procedures.

Since the user teams have been relatively large (up to 15 employees), the efficiency of the modeling work would have deteriorated if the entire user team would have participated in the modeling work. Thus, so-called "Power Teams" were created which were composed of two members of the user team and one method expert. Multiple power teams serviced one user team each. The total user team only met in coordination sessions in order to discuss the to-be processes presented by the power teams and to agree with them on further actions.

The Staff Committee and the departments Quality Management and Controlling had been involved in all decisions on the to-be processes.

Figure 2.3 summarizes the project organization during to-be modeling.

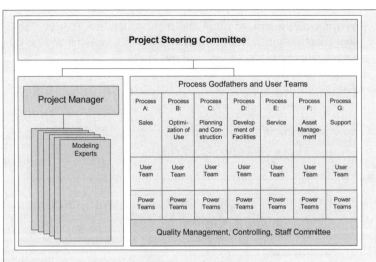

Fig. 2.3. Project organization during to-be modeling

DeTe Immobilien followed the new process structure and changed their organization. The introduction of this new organizational structure required the highest degree of conversion and integration competence.[25] In order to prevent each and every difficulty during decision finding (and there have been some problems) from being reported to the PSC, a decision level was inserted between the user teams who were created in accordance with the new organizational structure and the Project Steering Committee. This Project Decision Team was composed of the leaders of the six user teams – every leader was also a representative of management at the same time – and of the Organization & DP manager acting as leader of this project decision team.

... during organizational re-design

Project Decision Team

The user teams consisted of user experts from the head office, user experts from the branch offices, a modeling method expert, an executive in charge with the organization and one representative of the corporate staff committee. Every team was steered by the representative of one member of management.

The idea to install a project decision team had proven overall to be right. However, the user teams did not transfer all disputable decisions to this committee, but rather tried to solve the discrepancies in bilateral discussions between the user teams involved, in light of their higher practical know-how and because of their smaller team sizes, which made decision procedures more effective. In line with the progressing work on the new organizational structure, this direct decision finding, which followed the subsidi-

---

[25] See „Syndromes" in Chap. 2.5.

ary principle, improved more and more so that the Project Decision Team only needed to sign the prepared decisions and could
concentrate exclusively on critical questions.

Figure 2.4 shows the project organization during the definition
phase of the new organization.

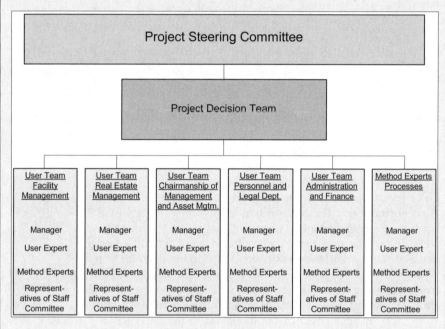

Fig. 2.4. Project organization during organizational structure design

Finally, the conversion phase itself had a new organizational form.
In accordance with the communication of the goals and measures
that are critical for the progress of the total project, a structure for
the implementation phase is required which enables rapid distribution of information and a quick response to questions without
overloading the central units. This requires observance of the geographical aspects as well as the integration of all management levels.

*... during
implement-
ation*

In order to support the implementation of the new organization,
special conversion teams were created for every branch office, as
shown in Figure 2.5. These branch conversion teams were responsible for the regional conversion and to maintain regular contact
with the head office. These teams were steered by the related
branch office manager and included three major groups of mem-

Conversion

bers: the conversion team for organization, the conversion team for communication and the conversion team for personnel. The conversion team for the organization defined the developed process parameters based on the individual regional situations. Simultaneously, the conversion team for communication took care of the provisions for technical and organizational information. The conversion team for personnel dealt with questionable items regarding the changed assignment of personnel and positions for personnel.

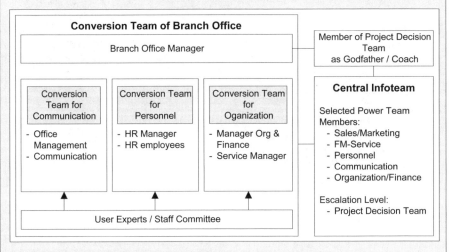

Fig. 2.5. Project organization during implementation

All teams were composed of user experts from different business areas, as well as members of the staff committee. Every regional conversion team was assigned a member of the project decision team as godfather / coach. This coach had to support the conversion and represented the team as the personnel contact to the central infoteam. This infoteam consisted of selected members from the power teams who had created the new processes and who were competent to answer subject-related questions. In case of major problems, the project decision team could be addressed as escalation level. The interaction of the individual teams is shown in Figure 2.6.

Coach and central infoteam

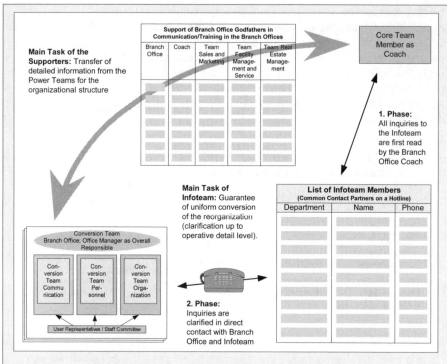

Fig. 2.6. Interaction of the teams during implementation

Constance in project organization

The different phases of a project go in line with different requirements in regard to the structure of the project organization. This, however, should not lead to a completely new secondary organization[26] for every project phase. Instead, the secondary organization should be designed from the very beginning to achieve an integration of the project phase through the constant availability of essential project members.

Embedding of departments

In the early phases of a reorganization project, the number of people involved (also the early embedding of employees from the departments) is relatively low, while the concrete conversion requires that absolutely all employees of the company are made familiar with the new processes.

---

[26] A 'secondary' organization is to the dedicated temporary organizational structure of the project, while the 'primary' organization is represented by the organizational chart of an enterprise.

2.4
Project controlling

Project controlling should be established as an independent function in the project management. Project controlling executes the
verification and controlling function during the processing of a
project in order to back up the defined performance and formal
goals. The task of controlling is also to prevent activities in the
project from occurring that are not directed towards the project objectives.

Tasks of
project
controlling

The effects of the project work will normally lead to an intru-
sion into the daily operations of current corporate procedures and
therefore often cause an open or hidden resistance. Those responsible for operation have to be convinced of the targets of the project and of the advantages to be achieved.

Project controlling is an important linking element to management, to the process owners, and to the support functions of the
company. In addition, it has a reconciliatory function for other current projects in the organization, in order to counteract redundant
developments. A continuous comparison of the objectives of the
project is vital, especially where projects with a major impact or
long-term projects are concerned. This way, the effectiveness of
the project is permanently verified and corrective measures can be
taken, if required.

Project
controlling as
linking
element

A controlling of the performance goals can often be executed
only after completion of a project task. In the case of process models, it is only possible to state whether or not the models are syn-
tactically correct according to the modeling conventions, and
whether or not the desired degree of detailing has been reached, after the modeling complex has been presented completely. Process
improvements, in particular, become measurable in an "effective
operation" only. Because of the relatively late practical test of
process improvements, special requirements have to be indicated
to project controlling in an early phase of the modeling project.
The created business process framework or the modeling conventions should be thoroughly verified early on since they essentially
influence the quality of the models to be created.

Practical and high quality models require thorough, continuous,
and professionally conducted discussion and an agreement among
the operative users in the creation phase as well as in discussions
of their visualizations of how to optimize the process flows. The
absence of a consensus or deviations from defined goals and standards force an agreement to be reached before the work is continued. Adjustments, working late, calling on additional specialists or
even a basic reorganization of the project may be required as well.

Collection of
process
models and
agreement

Processes and strategy

The ideas and concepts entered in the project have to be made compatible with the strategic corporate goals, especially with the organizational strategy or organizational principles of the enterprise. In addition, the increasing integration with the value chains of the customers and suppliers force a high degree of attention to related interfaces.

Time and cost control

The time and cost control is done by a periodical comparison of planned nominal parameters with the actual parameters. For the parameters to be valid for controlling they have to be thoroughly reconciled beforehand by the participants. In a "Feed-back - Feedforward" procedure, any deviations in the parameters have to be brought back in line with the defined goals.

Project management software

The comparison of parameters can be supported and secured by project management software. These tools are also efficient in supporting the planning of deadlines, the direct and indirect project costs, the utilization of necessary resources, the entry and tracing of other project-related data, as well as the identification and visualization of deviations between nominal and actual values. The methodic continuity is preferably reached by describing the essential project steps in a modeling technology that is in conformity with the process models. The use of the same modeling tool will enable branching out from the individual project steps directly to

Methodic continuity

the process models. In particular, it would be useful if these process models could be exported into the project management software for project description without major manual intervention. The project management software could then execute more clearly differentiated evaluations (e.g. calculation of the critical path, load of resources) than would be possible by the modeling software. Figure 2.7 gives an example for the interaction between the ARIS-Toolset and MS-Project.

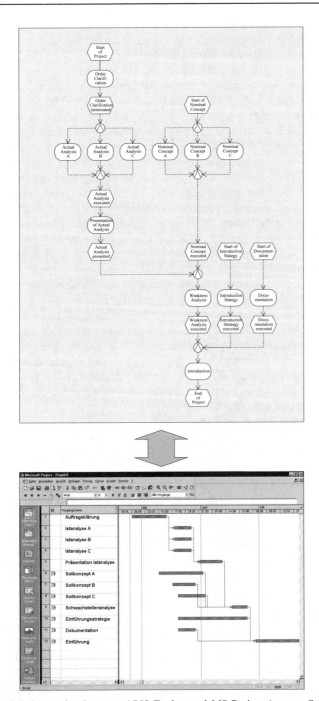

Fig. 2.7. Interaction between ARIS-Toolset and MS-Project (source: Scheer (1999), page 15)

Deviation from project goals

Project controlling within a project minimizes the friction losses, decreases the costs, and supports the orientation toward the targets. In practical project work, however, it will normally emerge that the planned goals cannot be reached completely in spite of all the sophisticated organizational and steering actions. The reasons, in addition to planning errors, are lack of understanding or lack of experience in the effects that result from project changes and which impact the actual processes. Doubts about the project are mixed with the personal fear of change. Normally, project management has to deal with a number of surprising smaller and larger interferences that are often hard to control. Often, overlapping projects suffer from the promised availability of user experts by the operative departments. The reason is to be found in poor planning of the involved department and a missing "binding obligation" for

Temporary availability of user departments

one's own personal participation in the project. If overlapping projects are pending in a company which require the cooperation of the user departments, then the necessary cumulated resources have to be calculated separately in the specific cost center planning of the company, otherwise the logical consequence would be a falling behind in deadlines and deviations from the projected costs.

A good project organization must always be prepared to either adapt the involvement of its required personnel in accordance with the altered capacity offer from the user departments or to vary its project activities to avoid an interference with the project progress by a lack of capacity.

Possible adaptations are:

Adaptation of capacities

- Redistribution of planned activities and movement of non-time critical tasks to the future (e.g., completion or detailing of process models).
- Temporary engagement of other internal or external project members (the additional "introduction" and the missing special know-how will cause an increase in costs).
- Improved rationalization in the settlement of tasks (e.g., stronger moderation and execution of project meetings).
- Reduction of demand in the settlement of tasks (e.g., only selective check of the syntax quality of the process models which have been created by experienced method experts).

Identification of moveable activities

The resource binding project activities shall be identified in such a way that it is clearly recognizable whether or not they can be moved. For these moveable tasks a strict and agreed upon time schedule is reasonable. An approach that can often be seen in practice is the refusal of project meetings in order to gain capacity. These meetings, however, serve to exchange the required know-how in critical phases. They should remain untouched, if possible,

but have to be executed in an efficient way, of course, i.e. by involving only the employees who are absolutely necessary.

An essential feature of project controlling is the consequent monitoring of the milestones and of those activities that relate to the critical path. Tasks that are not on the critical path need a reasonably timely synchronization as well as an efficient handling of previously allocated time buffers. The time buffers should be assigned, if possible, and should not be consumed by individual actions without coordination. A project progress report in short intervals, together with an early warning system, will help to recognize obstacles and to take counteractions in time.

Critical path

Time buffers

If the planned project costs are surpassed at any point during the project execution, then the remaining costs have to be investigated for possible reductions, in order to keep the overall costs as planned. Measures have to be initiated to reduce the costs, for example, through the use of more cost-effective resources or – in the extreme case – the omission of partial tasks of the project. Project controlling has the task of guaranteeing the backup of the total feasibility by a permanent "Stop-or-Go procedure".

Cost reduction

In modeling projects it has been proven practice to allocate responsibility for part of the prepared modeling activities to external contractors. These contractors must have undertaken basic related training. The personnel costs for those tasks in modeling projects could be considerably reduced by creating an "Intelligent Back-Office".

Intelligent Back-office

Project controlling must, however, not concentrate on the costs exclusively. Additional options for improvements that result in the course of the project have to be investigated for potential success and feasibility. This includes, for example, identifying previously unknown shortcomings thorough and comprehensive process modeling, whose elimination would be of major value to the company. Such additive improvements justify higher project costs when the total cost-benefit ratio is positive. Principally, project controlling has the task of rating the effects resulting from a project, under the economic point of view. The related surplus costs should not be an obstacle unless they blow the financial budget completely.

Value added

Financial reserves for those additional activities have to be provided for in the financial plan for a project. They must, however, be released only in the course of the project progress, and only if the operational managers accept that real improvements are likely to result from the additional activities.

Financial buffers

The task of project controlling has to be executed within the scope of the project organization, under consideration of the standard controlling functions of the company. Within this scope, the

Distribution of controlling tasks

Tasks of project manager

Tasks of Project Steering Committee

Quantity of projected advantages

Quality reporting

Project controlling at DeTe Immobilien

project manager and his team should not take over the total controlling function of the project. Superordinated controlling tasks are executed by the PSC. The PSC can nominate an independent controller as an additional entity that supervises the economic approach and the maintenance of monetary goals.

The project manager periodically checks the project progress in short intervals, sometimes even on a daily basis, and agrees on it with the workgroup, while the PSC is informed by the project manager in longer time intervals and is involved in the decision process. The PSC checks the milestones of the overlapping working packages as well as the development of costs and outcomes. The PSC makes the necessary decisions in regard to an adaptation to the project goals and to the configuration of the project.

Often, problems arise when the PSC has to verify actual project contents since the members of the PSC are not able to check the performance and quality in detail. The PSC can only execute spot checks. Therefore, the PSC needs a project manager who is experienced in this particular area as well as in the leadership of people. In addition, the PSC has to be composed of members who are experienced, far seeing, and competent.

The potential conflict often lies in the fact that the PSC wants to obtain precise information about a change of important control indicators at an early stage (e.g. detailed cost saving statements, percentages of improvements in the actual processing time or increased productivity parameters). This, however, can only be realized in the project runtime to a limited extent, and is restricted to the basics. Providing extra detail will not lead to better results. More important is the understanding of necessary changes, the recognition of potential improvements and their later consequential realization. Therefore, in certain cases, it is better to prefer a reporting on quality than to estimate the improvements by quantity where the anticipated results are subject to many uncertain variables.

A recommendation for the members of the PSC is to participate in the information- and decision sessions during project work in order to update their knowledge, or to be permanently informed by using the project management software.

DeTe Immobilien executed the project controlling in personnel in conjunction with the total project management function. Under the control and decision-making of a dedicated PSC, the typical controlling tasks were transferred to the project manager who executed them. This, for example, included the administration and control of the project budgets, the allocation of personnel in regard to time and costs, the control of the progress of the project, the

backup of the economic objectives, as well as quality control of the modeled processes. Periodic PSC sessions were held where the project manager reported the project status to the members partici- *Project* pating within the responsible management and staff committee. At *manager* the same time, actual problems were discussed and decisions were made on how to proceed with the project. In these target-oriented PSC sessions, the members of the committee executed the desired auditing function in addition to their control function. Important decisions, among others, were the modification of the project or- *PSC* ganization in the individual phases of the project, and the increase in user experts for the project in order to keep the defined dead-lines, but also to ensure the provision of the financial means for a sufficient participation of external experts.

In addition, upon request, the PSC members received updated *Special* information about the modeled processes outside the normal PSC *information* sessions via special information events. For this purpose, the proc- *events* ess models were edited and presented in different ways for each recipient group according to the specific requirements of the groups, since; for example, the management needed information different to the staff committee.

The comparison of projected costs with the projected profit at DeTe Immobilien during project work in the desired reporting *Project profit* format was not without problems. While the projected profit in the *vs. project* as-is modeling could be demonstrated by a list of detected short- *costs* comings, the to-be modeling enabled a monetary quantification only with difficulties. The to-be models could only show the po-tential profit. Approaches for improvement had been outlined, but the pending problems on the way to this profit remained unsolved. The theory was clear, but the practice of the new process-oriented organization was not yet detailed. The transparency of the total process model that had been created for the first time represented the basis and start-up of a complete process-oriented reorganiza-tion and a tailored new process communication in the company. DeTe Immobilien faced the necessity of achieving a new quality of performance in the entire company. All employees were involved, from the service engineer up to the board of directors.

DeTe Immobilien had the chance to optimize the processes based on a theoretically founded, transparent, traceable and practi-cable basis. The achievable success could, however, not be proven *Quality* in monetary terms, since success can only be visualized by control- *profit* ling after a new organizational structure is implemented and the employees are familiar with the new processes and responsibilities.

In the to-be modeling phase, all participants were expected to carry their activities on to the end. The business vision, the pres-sure of competition, and the expectations of the shareholders,

Quantity profit

however, were unable to cause a deviation from the selected way. After implementation of the completely restructured company in the head office and in all twelve operative branch offices, the benefits of process modeling and process orientation have been obvious for DeTe Immobilien. The improved productivity freed personnel capacity in favor of other tasks. The process costs became more competitive because of shorter processing times and increased quality, and led to a higher customer satisfaction. Now, all process levels in the company are clear, and further improvements will be realized faster and with more security.

DeTe Immobilien has executed a complete reorganization using the process modeling approach. It has received the DIN ISO 9001-certificate for the total company, and continues to work on the process orientation. The following improvements have been achieved:

Achieved improvements

- Optimization of the disposition processes. This led to an improvement of the disponent:service engineer ratio from 1:6 to 1:20.
- Considerable decrease of external services, and therefore a better balance of the own staff.
- Standardization of material procurement processes and clearly regulated and simplified competencies. The new processes enable the employees to respond to quickly changing situations in a fast and flexible manner.
- Elimination of redundant work and errors through clear and unique definition and documentation of organizational interfaces in the order management and invoicing process, making this process faster and, therefore, more cost-effective.
- Significant increase in customer satisfaction through unique determination of contact partners and by providing them with the necessary information. Upon request, customers get the required information about the status of their orders quicker and more reliably.
- The first and structured determination of the total sales processes supports DeTe Immobilien in their strategic goal to develop the group-external market.

2.5
Critical success factors

Employee syndromes

In spite of thorough planning and consideration of all organizational facts it may happen that a process modeling project does not lead to the desired success. Various syndromes with similar characteristics often cause this.

The "Not with me" – Syndrome

The inertia in some institutes is astonishing. Employees reject any changes and are skeptical about modifications. Since process modeling projects almost always lead to changes in the workplace – independent of the intended purpose – be it in the assignment of tasks, in the involved applications, or in the workflow, the employees regard the project with mistrust from the very beginning. They build an inner wall, refuse to give explanations, retain all information, point to others, do not transfer their own proposals for improvement, in short – they practice a silent boycott. Their offending attitude makes it difficult for project teams to push improvements and to accomplish them. The employees create a climate which is characterized by destruction, and often the members of other user departments, who had a more neutral or even positive attitude up to that time, are negatively influenced.

Readiness for changes

What to do: Since the problem is of a psychological nature more so than being a technical problem, the project teams need a great deal of empathy and conviction. Individual discussions have to highlight the positive aspects of the planned changes. This is especially important with the resistant employees and, sometimes, a clarifying talk is required.

The "Not invented here" – Syndrome

Frequently, the "Not with me" and "Not invented here" syndromes go hand in hand with changes which are transferred to the user department from outside. Changes from outside often have a lower chance of being accepted (and therefore to be converted) than those initiated by the user department itself. Therefore, it is important to give the employees a feeling that the planned changes are not a command.

What to do: The employees of the user departments should be involved in the idea-finding process, and the changes should be planned with them in workshops (as oppose to "presentation" of ideas that have exclusively been worked out by the core project teams).

Common creation of ideas

The "Do it yourself" – Syndrome

It is difficult to achieve success in a process-modeling project when the management – after having initiated the project – no longer have sufficient involvement in the project. Handing over the project to the project group must not be interpreted as management withdrawing from responsibility for the decisions to be made and for the conversion of the project. In order to increase the

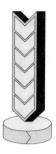

likelihood of success it is necessary for the management to maintain identification with the project, to provide a clear statement about the project to the employees, and to make dedicated decisions about changes. A delegation to the project group without the inclusion of take-over responsibility and without clear technical decisions from management, make it easy for the "protectors" and "objectors", for the "submarine drivers", and for the "we have never done this" people, as well as the "not-with-me" protesters, to gain dominance without representing the opinions of the majority of employees.

Visual engagement of the management of the company

What to do: Management has to show an obvious commitment by strong and convincing leadership as well as personal involvement in critical items and controversial opinions. Management also has to support the project group to make essential decisions, and to follow defined guidelines and policies.

The "Let's start immediately" – Syndrome

It is good when a project is started with commitment and conviction. But it is not good when over eagerness leads to a shifting of essential points in the beginning without having done the necessary reflection and preparative work. Preparatory work in establishing methodology and the structuring of the problematic areas helps to avoid later unnecessary work, which results from not thinking ahead.

Thorough preparation

What to do: The target has to be clearly defined, the scope of the project must be fixed, the method has to be selected, all project members have to master the method – via training, if applicable –, the process limits have to be determined (where does one process end and where does the next process begin?). These actions can be perfectly supported by a business process framework.[27]

The "Start and wait and see" – Syndrome

As with many other projects, the process modeling project is also subject to unforeseeable factors in terms of the required time and capacities. This, however, must not lead to a "start and wait and see" attitude. A defined time frame, as well as clear expectations for the achievement of goals, is necessary in order to give the project the required intensity. A project plan with time-and contents-related milestones is indispensable. This project plan has to be subject to strong supervision. Of course, it is very difficult to estimate times when the scope of work is not yet defined, but, on the other

---

[27] See Chapter 4.7.

hand, the deadlines can be kept by permanently adapting the time and intensity of the project team. It is a platitude that a project can only be executed successfully if it is characterized by a certain pressure.

What to do: Define clear deadlines, set milestones, create a realistic but challenging project plan, and execute a detailed project controlling.

Precise defaults

The "No time" – Syndrome

It is not unusual for the core project team to be freed from other tasks and therefore work with intense commitment and be involved in a great deal of time-consuming work. This work, however, can only be successful when all employees outside the core team take enough time to incorporate their know-how into the related project. Since this is not their primary task and, in addition, prevents them from carrying out their "important" activities, the time budgeted for and assigned to the project by the related employees is often extraordinarily limited. If information and the ideas of the employees with the related know-how are transferred to the core team, the core team could save considerable time.

What to do: The liable parties of the company are obliged to implement their good ideas. They have the know-how that enables them to propose improvements (and they have the power of influence to convert them). "No time" must be eliminated from the active vocabulary of these liable parties. In order to achieve this, the reliable parties have to be motivated to spend their valuable time in favor of the project and to accept, temporarily, a very high additional work load. In addition, they have to be freed from, at least partially, tasks from their everyday business.

Active time management

The "I don't care" – Syndrome

Project members have to contend with considerable resistance, be it because certain employees are afraid to lose their jobs, be it because of pure opposition ("not with me"), or be it because of disinterest or lack of commitment ("no time"). It would be fatal if the project members withdrew because of this high pressure and ultimately disregard their convictions and ideas for improvements. This would lead to a demotivation and in consequence thereof to a diminished convincing force for the project members. Since they are the driving forces in the projects, a weakening of this "drive" would cause these key people to lose their power.

What to do: Support project members by management, select persons with strong self-motivation and high convincing power for the project group.

Motivated project team members

*Conversion
competence*

The "Analysis / Paralysis" – Syndrome

Many companies not only fail to recognize good ideas, outstanding analysis, and excellent proposals for improvement, but they also suffer from a lack of competency for conversion. After presenting a proposal to reorganize the information system within the scope of a process modeling project, this proposal has to be converted into a finite (short) time. Even in the case of smaller changes of external variables, the conversion plan should be consequently processed. Nothing can paralyze a company more than an analysis that is followed by a new analysis and again by another analysis – for whatever reason.

What to do: Executing 80 percent of an improvement (measured from the optimized theoretical improvement) is still better than waiting for 100 percent fulfillment.

2.6
Checklist

*What to
observe!*

Project goals

- Set singular and practical operational goals in terms of performance, deadline, and costs of the project. Plan thoroughly and realistically. ☑

- Plan realistic milestones while considering resources available (especially human resources). Observe the effects on the costs when accelerating the project runtime. ☑

Project organization

- Adapt the project organization to the requirements of the different phases of the project, and secure the continuity in the "core area". ☑

- Engage highly qualified people for the Project Steering Committee, especially when a reorganization project is concerned. ☑

- Find the know-how carriers in your company and involve them in the project team. Make sure these employees are available since they are just the ones who are tightly bound by their operative business. ☑

- Involve the staff committee in the project team and in ☑ the Process Steering Committee at an early stage. This will allow identification of potential conflicts and their subsequent elimination.

- Identify the symptoms of the employees in as early a ☑ stage as possible. Respond to different concerns in a determined way in order to secure the success of the project.

Project controlling

- Don't anchor all controlling functions to the project ☑ manager. The PSC has to take over superordinated audit and control functions.

- Check the project status at periodic intervals for per- ☑ formance, deadlines and costs. Take efficient and effective corrective actions in case of deviations from the planned goals.

- Not all potential benefits can be calculated in monetary ☑ terms. Present potential benefits in early phases of the project by quality, and keep on tracing their conversion consequently.

Preparation of Process Modeling

Michael Rosemann

3.1
Relevant perspectives

3.1.1
Multi-perspective process modeling

The quality of conceptual models has been the subject of the academic research[28] for a long time. Comprehensive work has been conducted on the quality of data models[29], and, in recent years, related work analyzed quality issues related to process models[30] as well. Even so, a common consensus about the essential quality features of these models has not yet been achieved. Reports about the practical use of corporate-wide data models[31] and their insufficient benefits prove that significant modeling initiatives carry an economic risk. Therefore, process models have to be evaluated using the same factors apply as for any other product, i.e. time, costs, and quality. In particular, the errors made in the 80's, when company-wide data models were often created without a significant purpose in mind, need to be avoided when creating process models.

Quality of conceptual models

In academic discussions about the quality of a model, the characteristics of the model layout dominates[32], i.e. the minimization of the model surface or the average lengths of the connecting lines. Different research efforts have been made in developing elaborate approaches focused on the generation of model layouts, in order to improve the quality of models. This, however, assumes that the model quality relies solely on the characteristics of the model it-

[28] See Batini, Furlani, Nardelli (1985); Lindland, Sindre, Sølvberg (1994).
[29] See Moody, Shanks (1994); Maier (1996); Moody, Shanks (1998).
[30] See Rosemann (1996a).
[31] See Gerard (1993).
[32] See Tamassia, Di Battisti, Batini (1988); Brandenburg, Jünger, Mutzel (1997).

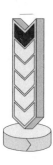

Criticism of existing approaches

self, independent of the model user or the modeling purpose In addition to criticism of the restricted view of factors which determine the model quality, many theoretical approaches can be criticized as well, in that they are analyzing "pen-and-pencil-models," which means that they take insufficient account of the far-reaching functionality of available modeling tools and their influence on the model quality.

In the following, the term "multi-perspective information modeling" will be explained. This approach requires the use of powerful modeling tools and a quality understanding as it has been developed in the Total Quality Management (TQM) approach.[33] This means that the quality of a product is not determined primarily by objectively measurable product characteristics (e.g. the share of chocolate in cocoa or the number of nodes in a carpet), but by the

Fitness for use

"fitness for use" for individual product consumers and the related purpose of use. This means, if transferred to the process models, that an identification of the users and their purposes of use are required in order to judge the "fitness for use". It can be stated that the quality of a process model is higher, the better it supports the purpose of use from the viewpoint of the user of the model. A per-

Multi-perspective modeling

spective combines the requirements that result out a combined consideration of purpose and model user. Process modeling for different user groups or purposes is called "multi-perspective modeling".

In contrast to industrial consumer products, which are also subject to multiple perspectives (e.g., a car with different types of drivers), a comprehensive process model can be potentially recre-

Total model

ated for any individual perspective. In this case a perspective can be seen as a subset of the total model that consolidates all elements and relationships.

3.1.2
Potential perspectives on process models

In the following, a short description is given for typical purposes of process modeling, as well as for typical model users. Together, this equals the description of potential perspectives on process models. A detailed description of the purposes, such as organizational documentation, process-oriented reorganization, and continuous process management follows in the next chapters. Additional purposes are also discussed in chapter 10.

---

[33] See for example Ishikawa (1985); Oess (1985).

Possible purposes of process models

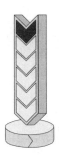

- Organization documentation
  While nearly every company maintains organizational charts, which are relatively easy to create and which are only modified to a minor extent in the course of time, just a few companies maintain up-to-date, comprehensive and relevant descriptions of their (business) processes. Existing process models concentrate usually on the purpose of on individual project and were created by individual employees independent of each other. The documents are not coordinated, neither in their contents nor in the modeling techniques, i.e. the way in which the contents are displayed (descriptive text or simple flowcharts prevail). The purpose of process modeling is to make the processes more transparent so as to communicate the processes more efficiently (i.e. for the training of employees). In addition, they can be used together with the organizational chart as input for job descriptions. This purpose requires that the sequence of functions to be executed as well as the related organizational units. The models should be intuitive because, in principle, every employee should understand the models. Since processes, by nature, are altered much more than an organizational chart, there is a greater need to keep the process models up to date.

  Availability of organizational charts…

  … but often insufficient documentation of business processes

- Process-oriented reorganization
  The most important reason for today's increased interest in business process models is the popularity of process-oriented reorganization – in the face of (revolutionary) business process reengineering[34], as well as in the sense of continuous (evolutionary) process improvement. Vital for this purpose are process models which help the involved managers and users to identify weaknesses of the process. At the same time, the models need to be sufficiently formal in order to allow a (partially) automated comparison of proposed new scenarios and as-is models. A demand for process models for this purpose is that they can be linked with the corporate goals. This requires an identification and hierarchical classification of corporate goals and their integration (as a new symbol) into the process models. This captures which processes support which goals.

  Business Process Reengineering

- Continuous process management[35]
  The continuation of the process-oriented reorganization is process change management, which means long-term planning, execution, and control of processes. Under the leadership of the process owner, the related process models have to be compared

  Process Change Management

---

[34] See Hammer, Champy (1993).
[35] See Chapter 9.

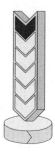

*Process
Controlling*

with the actual process execution. Any deviations must be in-vestigated to see if they result from an inadequate process model or from an ineffective or inefficient process execution. This forces the continuous process management to include process controlling. The attributes of the process model, such as the time parameters for each function can be updated automatically, if business activity monitoring solutions are implemented. Here, modern cost management approaches such as Activity-Based Costing be applied. The related tools should support normal controlling functions such as early warning systems or data aggregation. The models should meet the criteria of being clear and easy in terms of data maintenance.

- Certification pursuant to DIN ISO 9000[36]

*Documentati-
on is decisive*

The successful certification of a company pursuant to DIN ISO 9000ff is 50-80 percent based on the company maintaining a high quality documentation of its processes and quality assurance procedures. As such, organization models and business process models are required for this task. Furthermore, the use of sophisticated modeling tools provides a central repository, allowing for any changes to be made only once. As far as (further) required text documents are concerned, for example to create standard operating procedures for a quality management manual, the text can be generated from the graphical descriptions using one of the many available modeling tools. To achieve this particular information, for instance positions such as "Quality Manager", must, therefore, be integrated into the process models. Other conditions must also be met to ensure the documentation produced from the models is adequate. For example, it has to be proven that the processes are executed as specified in the documents. A basic prerequisite is also that the models can be understood intuitively without any special knowledge of the modeling technique. In addition, every model change has to be individually documented. Furthermore, a tight coupling of process models and organizational models has to secure an immediate and consistent follow-up of organizational changes. Finally, it is mandatory that every model and / or every model status be provided with a period of validity.

---

[36] See Chapter 10.1.

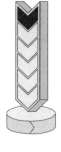

- Benchmarking[37]

  Benchmarking describes the approach of comparing the individual structure and performance with available other internal or external references. These referred reference values are supposed to represent the best practice or at least better practice. In the context of a process management project, this includes the comparison of the process structure as well as the actual process performance. Process models containing appropriate attributes are recommended for this purpose since they allow a comparison of the process parameters such as the process output. Any discrepancies can be taken as indicators for a need for deeper analysis. The pre-requisite for this purpose is the availability and comparability of the related processes which are used as benchmarks. In globalized organizations with different sites, branch offices, subsidiaries and the like, which have processes with identical structures, internal benchmarking can be institutionalized as well.

  Availability and comparability

- Knowledge management[38]

  Knowledge management aims for an increased transparency of the corporate resource knowledge. This covers the entire knowledge lifecycle including identification, acquisition, use, further development, and distribution of know-how. Process models which can be enriched in a versatile way are suitable for inclusion in operational matters and training courses (acquisition of know-how and know-how transfer). In this case, the process models have to be enriched with "knowledge" as input and output for functions. In addition, relationships between knowledge, organizational elements, and members of the organization have to be created and maintained. Finally, the core terminology of an organization has to be precisely defined and will contribute to the documentation and transfer of knowledge (refer to the technical term models, Chapter 3.5).

- Selection of ERP-Software[39]

  Enterprise Resource Planning-Software (ERP-Software) (e.g. from vendors such as SAP, Oracle, J.D. Edwards, PeopleSoft) can be defined as configurable off the shelf software that offers integrated business solutions for the core processes and support processes of a company. A comprehensive integration of data and functions supports different, typical functional areas such as procurement, operations management, warehousing, accounting, cost management, or human resource management. The functionality of this software is often documented in the form of

---

[37] See Chapter 10.6.
[38] See Chapter 10.5.
[39] See Chapter 10.2.

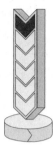

Reference process models

software-specific reference process models. This provides the opportunity to use these models within the ERP lifecycle covering system selection, implementation, use, upgrade and maintenance.[40] For the purpose of software selection an evaluation of the extent the reference model covers the requirements has to be conducted. The weighted scope of coverage (e.g. according to the importance of the process) can be regarded as one indicator for the suitability of the analyzed ERP software. Such an approach explicitly considers the possibly restricted sequence of functions (e.g. ordering without posting). Normally, the major problems with such a model comparison are the different degrees of details and the semantics of the models. These facts and also the different naming, structural and layout conventions render a comparison often impossible. Therefore, a judgment of the suitability of certain ERP software requires the generation of a model which is at least approximated to the software model, i.e. it should at least be the same modeling technique. The alternative would be to use the software-specific model as the starting point for the software evaluation.

Model comparison

- Model-based customizing
  Enterprise Systems require a comprehensive parameter-based configuration, which is also known as customizing. Reference models can provide valuable insights into the software functionality and can be an important input for related decisions. In most cases, the use of these models does not demand a deep technical understanding, so that representatives from the business units can be easily involved. Unfortunately, many reference models do not depict the actual configuration alternatives as they are using common modeling techniques. Thus, it is in most cases impossible to automatically configure the software based on modifications of the reference model.

- Software development
  The traditional application area for conceptual models is in the description of requirements for software to be developed as part of the "requirements engineering" phase. Here, mainly CASE (Computer Aided Software Engineering) tools are used. Process models that are designed with the specification of requirements are typically more formal and have well-defined relationships to other relevant models, such as data or object-oriented models.

Requirements Engineering

- Workflow management[41]
  A workflow is a (semi-)automated process, whose transitions of functions are under the control of a dedicated system, i.e. the

---

[40] See Rosemann (2000).
[41] See Chapter 10.4.

workflow management system. Process that are potential work-flow candidates are well-structured and have a sufficient number of instances per period. In contrast with process models for organizational purposes, workflow models have to be enriched by roles[42] (qualification, authorization), input and output data, including specific data structures and involved applications.[43] The relationship between organizational constructs (e. g. organizational unit, role) and function in workflow models is an "executes" relationship. In process models, this relationship can also have different semantics, e.g. "is responsible for". In comparison with business process models, workflow models normally have a finer granulation (degree of details) and forced by the demand to convert the model from the buildtime into the runtime a larger number of model attributes.

- Simulation
  Simulation allows a condensed analysis of the business process performance over time. It primarily supports the identification of weaknesses in a system which would not be revealed if only the model would be considered (such as insufficient capacity load, long idle times). As such, simulation facilitates a convenient comparison of different process scenarios. Another essential application area of process simulation is the calculation of the number of personnel required for a process depending on different scenarios which may occur. In order to be able to execute a simulation, frequencies, probabilities, processing times, cost information, as well as the availability of resources have to be specified in the simulation model. Similar to workflow management, the simulation of an instanciable model requires a certain degree of detailing as well.

Identification of weaknesses

The above listed purposes of process modeling obviously force the process models to meet certain different requirements in terms of contents and methodology. With respect to contents, the requirements differ in the related model components. Workflow models need, for example, a specification of the input and output data. This specification would not be required for process models which are designed for benchmarking purposes. Process models for benchmarking, instead, require a comparison of data. Such a data comparison, however, is not mandatory for workflow management. With respect to methodology, differences occur when the deviating requirements cannot be covered by one modeling tech-

Requirements

Contents view

Methodology view

---

[42] Here, the term role is used in the context of workflow management (see Chapter 10). This role does not match the role in the later discussed organizational modeling (see Chapter 7).

[43] See also Dehnert (2003).

nique. The models for the organization's documentation, the process-oriented reorganization, or continuous business process management need to be very intuitive. In order to achieve this clarity, self-explanatory icons are often used while process models for requirements engineering need to be assigned precise attributes and mandatory reference points to the corresponding model types (e.g. data or object-oriented models).

Figure 3.1 summarizes the discussed purposes of process modeling.

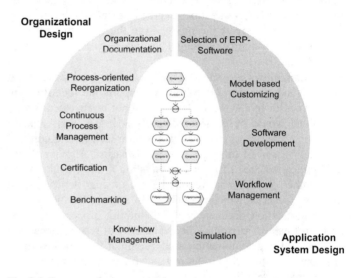

Fig. 3.1. Purposes of process models

Possible designers and users of process models

In addition to the purpose of modeling, the relevant requirements of the enterprise-wide process models are determined by those involved in the modeling activities. Two features are especially relevant for the determination of the model design: The methodological competence and the position of the involved users.

Methodological competence

The methodological competence describes the capability to create or understand process models using a specific modeling technique. The higher this competence the shorter is the time to create a model of a defined quality. Obviously, the methodological competence has two components, the actual and current competence, and the assumed competence in the future. Many modeling projects, however, suffer from the mistake of being exclusively based on the

current competence. The consequence is a correspondingly restricted selection of modeling techniques and tools. Often it can be
observed that established tools and techniques are used, because
the modelers are familiar with the techniques. These tools and
techniques might, however, not support the current purpose of
modeling very well. The modeling tool and technique to be selected should not be overly restricted by the methodological competence, but should preferably result from the relevant perspectives. The lower the methodological competence of the model
users is the greater is the importance of the model clarity.[44] Only, if
the methodological competence of the employees is very heterogeneous and if the models are used for very different purposes
(e. g. organizational documentation and requirements engineering),
may it be reasonable to actually use different but integrated techniques.

Furthermore, the perspective is characterized by the position of
the person involved. An employee who processes travel expenses
will outline the travel expense process differently than the re-
questor. The different contents chosen to be included in a model *Position*
may be intended and help new personnel in expense accounting or
serve to introduce new personnel to the handling of travel requests.
This is equally valid for the processes perceived by a medical doc-
tor and his patient, or for the interaction between customer and
supplier. These differences, however, can also be unintended. In
this case it is advantageous to follow the business process without
taking a special (position-dependent) perspective and to work on
the process model under participation of all involved organizational units. From this process model, position-specific perspectives can be derived at a later date.

3.1.3
Identification of the relevant perspectives

The relevant objectives of the modeling project have to be selected
from the potential purposes of modeling stated in the preceding
chapter. This requires holding early discussions with potential *Discussion*
stakeholders. These will be representatives from the senior man- *with*
agement, organizational management, IT management, quality ma- *stakeholders*
nagement, controlling, human resource management, as well as
representatives of the staff committee. Even though the modeling
techniques and tools should be determined and selected after the
perspectives have been identified, these discussions will be an iterative process. In practice it will be indispensable to present samples of process models in order to be able to make the general con-

---

[44] See Shanks (1997).

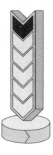

tribution of process models transparent. Various acceptance studies have to be conducted in order to get feedback for the popularity of alternative modeling techniques.

In addition, it is important to determine which project roles should be assigned to which persons, for example who shall design the models, who is in charge for quality assurance and who needs to understand the models? These decisions must take into account the subject and the application area of the processes. It is necessary to pay attention, however, that the competencies are not assigned too early or in a too comprehensive scope.

DeTe Immobilien first selected limited and well-known application areas (such as the messenger service), in order to give the involved employees an idea for the modeling approach and the practical use of process modeling. These examples were also used in the first project phase in discussions with the potential stakeholders. The related discussions were used to derive the perspectives which were of interest. They also helped to specify the requirements of the modeling method, the required configuration of the modeling tool, as well as the required specification of the model contents. Initially, the following purposes were regarded to be relevant:

- Documentation of the organization
- Process-oriented reorganization / continuous process management
- Certification

The following purposes were regarded to be relevant for the near future: Evaluation of certain ERP modules for suitability, workflow management, and simulation for personnel requirements planning.

3.1.4
Differentiation of perspectives

Contents and method criteria

After the relevant perspectives have been identified, the requirements of the process model can be consolidated. These requirements form the basis for the way in which the process models are created, in particular for the related modeling conventions. Each perspective can be seen as a subset of a more comprehensive process model. Individual perspectives can be provided by a combination of six different features. In the following, these features are explained in more detail.[45]

---

[45] See Rosemann (1996c); Rosemann (1998b).

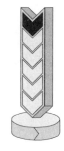

- Different layout conventions
The layout conventions for two perspectives differ from each other when the number and names of the model elements match but their graphical design (e. g. a rectangle instead of a hexagon)[46] or the topology (e. g. arrangement of the process flow from left to right instead of top-down) mismatch. In the latter case, algorithms for layout optimization[47] can be applied, which have to take the individual focal points of interest into account (such as the selection of the entity types to be arranged in the center of a comprehensive data model). Layout optimization is equally valid for esthetic criteria (e. g. minimum / maximum average length of edges, minimum overlay of edges, minimum surface at a given standard distance, transparent symmetries). Ensuring clarity of the model and / or modeling technique is the criterion for a formal model quality. This criterion considers that different model users have differing degrees of experience in modeling.

  Layout Optimization

- Different name conventions
The internationalization of existing modeling projects increases the necessity to maintain alternative terms in different languages (e. g. in a technical term model, refer to Chapter 3.5). The users of the models have to be grouped and assigned the relevant terms (e. g. company-internal, country-specific or software-specific vocabulary). The importance to maintain such synonyms decreases with the establishment of quasi standards (e.g. communication in English) or with the adaptation of popular software-specific terms (such as the SAP term "company code" instead of "company").

  Quasi Standards

- Different objects
When perspectives differ in regard to the objects used in a model, the differentiation of perspectives orients more towards the actual model contents. For workflow specifications, for example, manual functions with a high degree of detailing can often be regarded as irrelevant. Similarly, pure DP (Data Processing) activities (e. g. batch processing processes) can be omitted in a perspective that is to be used for organizational design. In this case, the perspectives are represented as different projections on one common process model. This form of a perspective differentiation is extremely expensive in its conversion to software since it requires the assignment of any individual object to one or more perspectives. On the other hand, it is exactly this feature which contributes to greater acceptance from users because it reduces the complexity of the model considerably.

[46] See Scheer (1998b), p. 76.
[47] See Brandenburg, Jünger, Mutzel (1997).

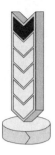

- Different object types
  Often, it will be possible to generalize the divergent require-ments of two perspectives in such a way that a relation can be established between the perspectives and the object types. In process models, for example, the purpose determines the re-quirements; i.e. models may or may not specify input and out-put data, or embedded organizational constructs. Correspond-ingly, the perspectives are represented as different projections on one common meta model. The differentiation between the perspectives is reflected in the modeling technique and in the object types used for modeling. This perspective differentiation is already supported by modeling tools which allow a configu-ration of the modeling techniques.
- Different use of modeling techniques
  In addition to the different relevance of individual model ele-ments, it is conceivable that there are also differences in the in-terpretation of the modeling rules. The motivation for the dif-ferent applications of the agreed modeling technique can be according to different weightings of quality criteria for the model; i.e. model clarity or correct syntax or the possibility of an automatic transformation of the model. An example is when an event-driven process chain (EPC) is modelled as a sequence of functions, rather than following the originally defined rules of (function - event - function). While the omission of the "event" element conflicts with the defined rules, it is often re-garded as being much clearer.[48]
- Different modeling techniques
  The highest individualization of perspectives occurs when a combination of different modeling techniques are tolerated. Such a differentiation of individual perspectives is necessary when the requirements of the relevant perspectives (e. g. calcu-lation of processing time, clarity) cannot be serviced by one single method. In order to master the redundancies which result from the coexistence of these methods, explicit relations be-tween the modeling techniques need to be set and formally de-fined. In actual fact, the specifications of most modeling tech-niques, (i.e. their meta-models), are not yet published in a formal format.

The discussed approaches to perspective differentiation clearly re-veal the growing complexity of modeling – the great number and variety of models as well as possible inconsistencies between them. On the other hand, the intention is to place the models in

---

[48]   See Scheer (1998a).

proximity to the customers and to increase the model quality. For every perspective to be developed, both the related added value (increased "fitness for use") and the corresponding expense have to be investigated. For each perspective, the benefits of creating, using and maintaining an individual perspective have to be evaluated. The most important forms of perspective differentiation place a high demand on the functionality of modeling tools.

Critical evaluation of the benefits of a perspective

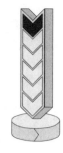

3.2
Relevant modeling techniques

The essential requirements of a modeling technique are based on the identification of the purposes and on the modelers or users involved in process modeling. In contrast with data models, no quasi-standard has yet been established for process modeling. In the following, typical requirements are listed for process modeling technique. The focus is on modeling for process documentation and process improvement.

- To clearly present the sequence of functions, including splits and joins. To allow different model hierarchies as well as to link process models on the same level via interfaces.
- To relate the process model to the data models (data for input and output), to the organization models, to functional decomposition diagrams and further relevant models.
- To define the modeling techniques in a sufficiently formal format in order to be able to provide at least a valuable basis solution for extended applications, such as simulation, software design or workflow management.
- Finally, it is vital that a tool is available which supports the modeling technique. In fact, the advantages of a modeling technique, as well as those of the modeling tool, will always be evaluated together.

Requirements

At this point, a comprehensive comparison of alternative process modeling techniques is not provided.[49] Here, the modeling technique that has been chosen within the scope of the DeTe Immobilien project will be outlined. The selected event-driven process chain (EPC)[50] was the only one that fulfilled the requirements at the time when the techniques were evaluated.

EPCs are directed graphs which use three basic elements for modeling the control flow.

---

[49] See Hess, Brecht (1996).
[50] See Scheer (1998c), pp. 125-127; Becker, Schütte (1996), pp. 55-57; Rosemann (1996a), pp. 64-66.

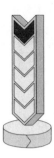

Functions

Events

Connector

- Functions represent activities. They are active nodes and transfer input and output data. Functions are executed by resources and can decide on the further process flow. In the graphs, they are identified by rectangles with soft rounded corners.
- An event depicts a flow-relevant status. In contrast to functions, events consume neither time nor costs. Four essential types of events can be distinguished as follows:
    - An event identifies a new process object (e. g. "Master data created", "Order created") or the final status (Delete) of an existing process object (e.g. "Invoice archived", "Project terminated"). Often, this is valid for the start and / or end events of a process.
    - An event refers to a change of an attribute (Update) of the process object (e. g. "Invoice verified", "Application is entered"). This does not have to be an IT-related event (e.g. "Customer called", "Truck arrived").
    - An event describes a certain point in time (e. g. "Due date for reminder").
    - An event stands for a change of the database which initiates a process (e. g. "Stock below reorder point", "Credit line surpassed").
- Events are presented within EPCs as hexagons and do not have the capability to make decision. They potentially fulfill two tasks. One task is the initialization of functions (initializing events, e. g. "Invoice has to be entered"). The second task is the documentation of a status that has been reached after one or more functions have been executed (reporting event, e. g. "Invoice is posted").
- Logical connectors are used to model splitting and joining points in process models. The connectors are distinguished as follows:
    - Conjunction (AND; "a and b") which is identified in process models by the "AND" symbol $\bigwedge$,
    - Disjunction (XOR; "a or b") which represents an exclusive OR and is identified by the symbol (XOR)
    - Adjunction (OR; "a or b or [a and b]"), which is represented as inclusive OR by the symbol $\bigvee$.

    Identical or different connectors can follow each other immediately. When a split corresponds with a corresponding join, the two operators have to match.

In standard EPCs only different types of nodes are allowed to be connected (here: functions and events). This means that one or more events initiate the execution of a function and then generate, in consequence of this execution of a function, one or more events. An event-operation exists, if multiple events are connected with one function. A functional operation exists if multiple functions are connected with one event.

In addition, the modeling rules of the EPCs contain two further rules.

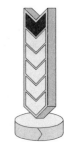

Notation rules

- Every process model must start and end with at least one event. This does not only assure that the start and end conditions of the process are specified, but this also corresponds with the real fact. Every function must be preceded by an initiator and every function must lead to a change in status. Because an EPC only describes a part of the process, process interfaces are used to connect the preceding or succeeding processes.
- An event must never be directly followed by an XOR or OR split since an event has no decision competence to determine the further flow of the process.

In addition to these basic notation rules and objects, EPCs can be enriched with a large number of additional objects. This leads to the extended event-driven process chain. Of special relevance are data, organizational units, application systems, and outcomes

- Data can be assigned to individual functions using an input and / or output operation (Create, Read, Update, Delete). These operations map the useable data of the process model. This data also helps to characterize the status of the event in more detail. For both purposes, the symbols of the corresponding data model (cluster, entity type, relationship type, attribute) are used.

Useable data

- The linking of functions with organizational units clarifies who has the task and competence to execute the function. In addition, other relationship types are maintained such as "has to be informed about" or "is involved in". This additional information allows identifying the organizational interfaces along the process or deriving process-oriented job descriptions.

Organizational units

- When the functions are processed automatically, this can be described by assigning an application system. For process analyses, media breaks can be detected relatively easily.

Application system

- For every function and for the processes in total, outcomes can be defined. These outcomes can be both products as well as services. The visualization of outcomes makes the overall contributions of functions and processes transparent.

Performance

The event-driven process chains are implemented in an integrated "multi-view" organizational framework, the Architecture of Inte-

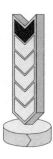

ARIS

grated Information Systems (ARIS[51]) which consists – in addition to the process view – of a data view, a functional view, an organizational view and an output view. Figure 3.2 gives an impression of how the EPCs are embedded in this architecture.

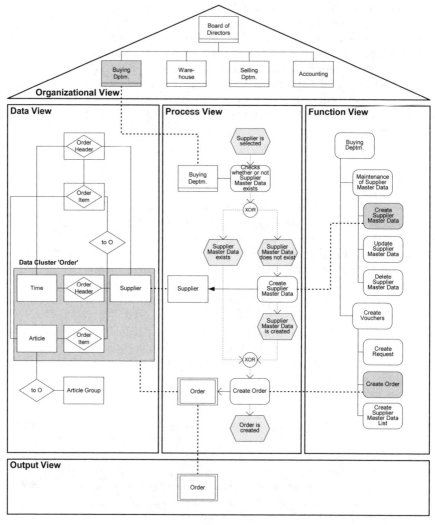

Fig. 3.2. Integration in the Architecture of Integrated Information Systems (ARIS)

---

[51]   See Scheer (1998c); Scheer (1998b).

In addition to the division of the architecture into five description views, ARIS is further defined by a subdivision into three description levels following the software lifecycle. These levels are: requirements specification, logical concept, and implementation. The process models which are discussed in the context of this book are assigned to the requirements engineering level.

Description levels

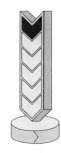

An example for a model on the second level is a workflow model which contains data structures. Finally, the models of the implementation level include detailed specifications (e.g. in SQL).

Of particular relevance for DeTe Immobilien was the process view. In addition to event-driven process chains, highly aggregated processes were designed, so-called value chain diagrams. These value chains contained exclusively functions – from left to right. These presentations focused also explicitly on the interfaces to the external customers. Each individual function in the value chain diagrams were described in detail in event-driven process chains.

Value chain

Even though the process models have been enhanced by objects depicting data, organizational units, and application systems, the independent maintenance of the other views was only of minor importance. Within the organizational view, the actual and the future organizational structure have been documented. In the data view, the IT department maintained data models as well as technical term models. No models for the functional view had been created, nor for the output view.

Relevant views

3.3
Relevant modeling tools

The identification of the relevant current and future perspectives, the purpose and the model user, as well as the selection of adequate modeling techniques, are followed by the selection of the modeling tools. As already outlined, this is typically not a sequential but an interactive process.

A modeling tool to be used for a comprehensive company-wide modeling project which supports the design of as-is and to-be models by modelers at various sites, should meet the following requirements:

- Support of selected modeling techniques and / or comparable techniques.

Requirements

- Adaptability to company-specific purposes (elimination of irrelevant functionalities; addition of new aspects).

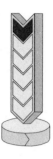

- Redundancy controlled model administration by a view-overlapping meta-model (concept) and an integrated database (system).
- Availability of relevant reference models in supported modeling technique.
- Interfaces to related add-on modules or third party solutions so that the project outcomes can be used for a variety of purposes (e.g. activity-based costing, simulation, workflow management, system development).
- Multi-user capability (simultaneous design of models and model access from different sites).
- Internet capability (conversion of models in HTML or JAVA format, navigation through the models using web browsers).
- Sophisticated user support by wizards, online-help and reporting functions.

ARIS-Toolset	These requirements could be fulfilled at the time of the project in major parts only by the ARIS-Toolset from IDS Scheer.[52] Thus, the ARIS-Toolset had been selected as a modeling tool. However, the desired administration of various perspectives is not yet possible in the desired form with this tool, nor with other (tools).

3.4
Modeling conventions

3.4.1
Guidelines of modeling

In many companies, the increasing necessity to come to an understanding of the business processes and to describe these business processes in a transparent manner, has lead to the co-existence of many modeling purposes, modeling techniques, modeling tools, model users and models. This

Increase in complexity

This welcome increased interest in process modeling, however, has resulted in a considerable increase in the complexity of the process models as well as process modeling. This complexity can be expressed by the variety of techniques, tools, modelers and users as well as by the quantity of the models. This makes it difficult

---

[52] Details of the selection of modeling tools are not discussed here. Refer for further information to Chrobrok, Tiemeyer (1996); Finkeißen, Forschner, Häge (1996); Buresch, Kirmair, Cerny (1997); Fank (1998).

to ensure consistent maintenance of the models, and their consequent acceptance.

These changed requirements of process modeling can no longer be satisfied by an academic discussion that focuses on syntactical issues. Instead, further recommendations for the design of comprehensive models are required. These recommendations should have a standardizing effect and help to manage the upcoming complexity. The consequential goal is to reduce and / or master the complexity of modeling. This problem motivated the development of the Guidelines of Modeling (GoM)[53] .

Mastering of complexity

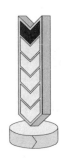

The terminology of the GoM orientates itself toward the generally accepted accounting guidelines. The aim of the GoM is to contribute to an increased quality of information models. Six guidelines define the structure and are regarded as the main quality criteria for information modeling. These guidelines are:

Intention of GoM

- Guideline of correctness
  An essential pre-requisite for a high-quality model is the actual correctness of the model, i.e. how well has the relevant part of the real-world been mapped into the model. This includes the described structure (e. g. the organizational hierarchy) as well as the described behavior (e. g. the processes). This so-called "semantic correctness" should be separated from the syntactic correctness which describes the correctness of the model in comparison with the underlying modeling technique and any company-specific rules.

Guideline of correctness

- Guideline of relevance
  This guideline relates to the already-discussed perspectives of models and their different requirements. On the one hand, a model has to document the relevant part of the real-world (e. g. cost rates in a process model for the purpose of activity-based costing). On the other hand, the model must not contain any irrelevant information.

Guideline of relevance

- Guideline of economic efficiency
  The purpose of this guideline is to ensure that the modeling activities remain in a reasonable cost-benefit ratio and will not develop into "l'art pour l'art". The efficient creation of models can be supported by the use of reference models or through the

Guideline of economic efficiency

---

[53] The Guidelines of Modeling were first explained in 1995 in Becker, Rosemann, Schütte (1995). Guidelines of process modeling are discussed in Rosemann (1996a). Additional specific modeling recommendations for data, organizational and functional modeling can be found in Becker, Schütte (1996). The GoM-Advisor, a tool to administer modeling recommendations with an integrated model for approaching the creation of models is presented by Becker, Ehlers, Schütte (1998). An overview of the GoM development is given in Rosemann (1998a).

practice of model re-use. In order to be able to judge whether or not the goals have been reached in this respect, a posting of the related costs and an assignment of the benefits achieved by this modeling needs to be performed. A task that is often extremely difficult in practice.

- Guideline of clarity

Guideline of clarity

This guideline takes the individual user of a model into account. A model should show an adequate degree of intuitive readability in consideration of the model user, i.e. the knowledge required to understand the model has to be determined. This guideline guarantees that a model is not only correct, relevant and potentially beneficial, but can actually be used.

- Guideline of comparability

Guideline of comparability

This principle guarantees the uniform application of the recommended modeling conventions within and between individual models. It simplifies considerably the consolidation of models in a comprehensive modeling project.

- Guideline of systematic design

Guideline of systematic design

Models always reflect only a partial aspect of a defined real world section (e. g. a focus on processes only, without related data). Consequently, well-defined interfaces to corresponding models are needed. This means that the input data in a process model should refer to a data model.

These six guidelines are not only discussed on this generic level, but can also be established for each view (e. g. guidelines for data modeling or process modeling). When GoM are designed for specific modeling techniques, modeling conventions to the greatest degree of detail can be derived based on this framework.

3.4.2
Intention of modeling conventions

The Guidelines of Modeling are applied in the form of modeling conventions (also: modeling standards). Modeling conventions

Reduction of variety

guarantee a uniform use of modeling techniques, and thereby increase the model quality by decreasing the variety in the model design. In this way, the models can be compared (e. g. as-is and to-be

Comparability of models

comparison). The reduction in inconsistencies can also result in a simplification of model-wide analyses (e. g., in which processes is an organizational unit involved?). The clarity of model evaluations is also improved by conventions which determine the attributes

Clearness of analyses

which have to be maintained. At the same time, explicit modeling rules reduce the uncertainties of the modelers in those cases where a certain degree of freedom is given (e. g. use of oval or rectangular organization symbols) or where detailed knowledge is required (e. g. selection of object types which can be used by an interface program to generate a workflow model). This accelerates the modeling process and reduces the number of later model adaptations.

Acceleration and simplification

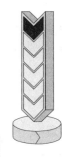

Modeling conventions do not define modeling techniques. They rather describe how a given modeling technique and modeling tool should be used in the context of a specific project.

DeTe Immobilien developed prior to the creation of the process models their modeling conventions. The structure followed the Guidelines of Modeling. The modeling conventions, which existed in the group already, were adapted to the perspectives identified in this project. Thereafter, the perspectives were individualized. The document outlining the conventions consisted of 70 pages and served the modelers as an operating manual for their work. Its details were continuously modified in the course of the project.

Conventions at DeTe Immobilien

The modeling conventions deemed to be relevant for DeTe Immobilien will be discussed in the following chapter. Extractions of the individual project modeling conventions can be taken from the appendix.

3.4.3
Subject of modeling conventions

By further specifying modeling techniques, object types and attribute types, modeling conventions restrict the usage of the selected modeling technique and tool. Conventions are used in the beginning of a modeling project to configure the modeling tool. In addition to providing guidance for the individual process modeling initiative, conventions also contain a specification of the user groups, the layout and naming conventions and a description of all responsibilities along the lifecycle of a model.

Restrictions

Configuration of tools

The sequence in which the modeling conventions should be built up corresponds to the sequence as described below.

- Modeling techniques
  The choice and definition of the modeling techniques to be used[54] is a basic decision. It is also the most far-reaching decision with respect to its impact on the conventions. The definition should be based on the relevant perspectives and should also state which modeling techniques will support these purposes in a form that can be understood by the user. The recommendation of alternative techniques (e. g. two approaches for process modeling) should be avoided unless both approaches can be automatically converted into each other, without major

---

[54] See Chapter 3.2.

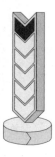

Process model

losses of contents, and unless their layouts differ significantly, i.e. address different user groups. A matrix should show the existing relationships between these techniques (e. g. "can be detailed by"). When possible, a process model should be developed that describes the sequence (in particular, top down or bottom up) in which the individual models have to be created for a particular subject area (e.g. mandatory creation of a value chain). It is essential in any case to take into account the statements of the interviewed users, modelers, creation date, and model status. The conventions should allow an explicit indication of text which can be freely formatted and to which notes can be added, if required (i.e. limitations regarding model completeness). This will also improve the tracing of model related decisions (e. g. "Process was created in adaptation of a software-specific reference model"). A matrix should show which perspective requires which modeling techniques.

- Object types

Selection of relevant object types

The determination of the modeling techniques to be used restricts the relevant object types within a model. Each modeling technique should be studied regarding the required set of object types with a focus on the (optional) object types. The available object types have to be critically discussed, for process modeling in particular, since the use of too many object types will considerably increase the complexity of the process models and simultaneously decrease their clarity. Normally, the relevant object types must be selected from alternatively available organizational and related constructs (among others, organizational unit, project team, position, role, employee, site). The questioning of an object type should always be the basis to determine whether or not an object type is used in multiple model techniques, and therefore contributes (or not) to the integration of models. The object list (which object type belongs to a modeling technique?) has to be completed by a where-used list (which object types are used in which modeling technique?). Subsequently, it has to be determined for each object type which attributes have to be maintained, while distinguishing between mandatory and optional attributes. This decision requires particular attention to the functions which are embedded in the processes, because the maintenance of time and cost attributes translates into diverse advantages (e. g. elaborated reports such as critical path analyses or determination of costs for a process throughput), but also into substantial modeling efforts and, in many cases, quantification problems. Attributes which are treated the same way in all object types (e. g. editor, interview partner, date), can be regulated in advance and be mandatory for

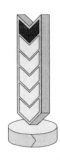

all object types. A matrix should then show which object types are used in which modeling techniques, and whether the object types are mandatory or optional. If required, this matrix should also be able to show the requirements of individual perspectives. Another matrix should show which object types can be further detailed by which modeling techniques.

- Roles of relationships
  The selection of the relevant object types is followed by the specification of the relevant roles of relationship types between the object types for any individual model. Through these relationship types, object types are brought into relation with each other (e. g. data and function). While these are obvious for many combinations (e. g., sequence of control flow), there are, in cases such as the relationship between organizational unit and functions, several possible so-called roles (e.g. "executes", "is responsible for", "has to be informed about").

The definition of the modeling standard must not be underestimated in its complexity, meanwhile keeping in mind the considerable functionality of the modeling tools. A modeling tool can easily provide more than 100 modeling techniques, more than 150 object types and more than 500 relationship types. *Complexity of conventions*

After the customizing of the modeling techniques has been finalized with the definition of the model-, object- and relationship types, these elements have to be put in relation to each other in respect to the different purposes and user groups. Subsequently, a complete cross product of conventions exists (purpose of use x user groups). Ideally, the modeling tool, which is used and correspondingly configured, should only provide the user with the relevant subset of modeling constructs that are needed after the user has been assigned to a user group in a login procedure and has specified the purpose of modeling. *Reference to perspectives*

In addition, layout conventions are an elementary part of modeling conventions. For each modeling technique, the relative spatial arrangement of the individual object types to each other has to be defined. In a top-down process model, where different cases and paths are depicted, the typical process path should be arranged more to the left, and the exception case more to the right. The length of the edges should be minimized and overlapping edges should be avoided. Precise guidance has to be provided regarding the maintenance of information such as the model title, model version, the embedding of freely formatted texts, logos or all other formatting such as margins, header and footer, font size etc. *Layout conventions*

The degrees of freedom of the modelers are further restricted by naming conventions which regulate, as far as possible, the designa- *Naming conventions*

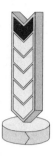

Degree of detailing

tion of the models as well as of the object and relationship types. The naming conventions which are mainly defining the terms to be used as well as the structure (e.g. "invoice entering" vs., "enter invoice"), should refer to standardized terminology. This can be maintained in a technical term model.[55]

Conventions which outline the agreed degree of detailing of the process models, are not an elementary part of the modeling standard, but always among the most challenging questions in a modeling project. Past projects indicated that there is practically no possibility of determining "objective" criteria for deciding on a reasonable degree of detailing of a process model. Reference points could be that a new function has to be modeled after every change of an organizational responsibility, or after every change of the supporting application system. Another example is that modeling has to be done on a level that enables a continuous description of the input and output data as data clusters, entity types or attributes;[56] or that objects are used, such as invoice or product. All these proposals, however, cannot cope with the heterogeneous requirements of a company-wide modeling project. Instead, the adequate degree of detailing is determined exclusively by the individual modeling purpose. The modeling team has to find this degree of detailing in a "trial-and-error" approach. The modeling team must further pay attention that the models created independently by different modelers show a comparable degree of detailing. This can be achieved by an early and intensive interaction between the members of the modeling team.

Modeling tool

The modeling tool should be configured as much as possible according to the defined modeling conventions and thus prevent the standards from being violated. This configuration is relatively simple when using so-called "model filters" which determine the modeling techniques, object and relation types, each with the related attribute types. On the other hand, the configuration for layout and naming conventions is more complicated to implement. After modeling, additional checks can be made by reports (e.g. constant maintenance of mandatory attributes, unless the maintenance is forced by the tool).

Model lifecycle

This also shows that further organizational rules are needed in order to guarantee the conformity of the created models with the modeling standard. This requires the definition of precise responsibility along the process model lifecycle. The lifecycle phases describe the different states of the model, which have to be main-

---

[55] See Chapter 3.5. A detailed discussion of naming conventions for process models, precisely for event-driven process chains, is given in Rosemann (1996a), pp. 187-189.

[56] See Hoffmann, Kirsch, Scheer (1993), pp. 13-15.

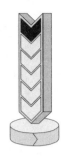

tained in the model attributes. Typical lifecycle phases are "in process", "released for approval", "approved", "checked for conventions", "(technically) integrated in database", "invalid", or "to be deleted".

A modeling standard should also regulate the technical responsibilities (creation of new databases or directories, user management, upgrades of the modeling tool) and assign these tasks certain
roles (system administrator, database administrator). In addition,
the procedure has to be specified as to how to integrate new models into the database.

Technical responsibility

Finally, it has to be noted that the discussed modeling conventions can be positioned on four levels:

Levels of conventions

- Reference-modeling conventions are comprehensive general conventions. An example would be the creation of models for certification purposes. These conventions are normally devel-oped by consulting companies, central departments in company groups or software houses who develop, manufacture and sell modeling tools.

Reference conventions

- Company-specific modeling conventions are modified reference modeling conventions or newly created conventions tailored to the needs of a company. In the case of modified conventions, modeling techniques, object-, relationship- and attribute types no longer needed are eliminated, and company-specific symbols added.

Company-specific conventions

- Project-specific modeling conventions are derived from the company-specific conventions. A company can have different conventions for different projects. In order to secure a maxi-mum compatibility of these conventions among each other, pro-ject-modeling conventions must always be defined as subsets of the company-specific conventions.

Project conventions

- Perspective-specific conventions are a subset of the company-specific conventions or even of the project-specific conventions. They contain the conventions which are relevant for a specific user group (e. g. modeling team), for a specific purpose (e. g. creation of a process model suitable for simulation) or for a per-spective, i.e. a combination of user group and purpose.

Perspective-specific conventions

Examples from the Convention Manual that has been created within the scope of the DeTe Immobilien project can be found in the appendix.

3.5
Technical term modeling

Modeling conventions essentially restrict the modeling techniques to be used but do not define semantics. Often, traditional glossaries are unable to take into account the company-wide agreed terminology, because they are not up to date, lack acceptance or are inconsistent. Traditionally, terminology is maintained as a body of text. This form does not harmonize well with storage media and editing. Technical term modeling[57] can provide an interesting alternative in the context of enterprise modeling. The concept of technical term modeling has close relationships with process- and data modeling. The integration of technical term modeling in a modeling tool can increase the consistency in the terminology used, and can increase the quality of data- and process models. In addition, technical term models are of special importance for knowledge management since they support the creation, comparison, communication, and continuous maintenance of subsets of the corporate knowledge, i.e. the meaning of relevant terms and their relation to each other.

Problem: Traditional glossary

3.5.1
Intention of technical term modeling

The diversity of terms in a company burdens the communication between the employees from different departments. This is the case, for example, when processes span technical and commercial functional areas and the related individual terminology is manifested in the related parts of the process. This results in a considerable complexity which is explained by the numerous and differing vocabularies used in a company.

Heterogenous terminology

The problem with the existing terminology and its numerous and varying – as well as unclear – vocabulary is also reflected in the data and process models and their designers, respectively. The heterogeneity of terminology makes it difficult, for example, to decide on one data model, to decide which entity types are relevant, how to name them or which semantic relations they have with each other. Process modeling is complex due to the identification of the relevant processes, the definition of the input and output of functions, and time-consuming in the later course of reconciliation ("model approval") of the created process models by the users from different departments. Consequently, the models often show lacks in quality, in that identical activities have different names

---

[57] See Kugeler, Rosemann (1998).

(synonyms). As a consequence, identical (redundant) activities remain undetected, or different activities are assigned the same names (homonyms).

A major task is to reduce the complexity (of the models) by eliminating as many linguistic errors as possible (synonyms, homonyms, vagueness, etc.[58]).

The prevailing methods for data modeling, however, are often not sufficient to come to a consensus on a definition of terms. More intuitive modeling techniques and terms are required to involve employees who do not have special knowledge of data modeling. Therefore, a technical term model is an efficient option to communicate with users because of their more intuitive nature. At the same time, a technical term model is sufficiently formal so that it can be used as a basis for data and process modeling. In addition, the technical term model supports the goal to collect and consolidate the relevant terms. The basic method has been described by Deutsche Telekom AG[59] and further developed by DeTe Immobilien.

Inferiority of data models

All vocabularies which are required to describe activities in conformity with the organization's targets and which are worthy of being defined – which includes the terms to be defined in context with a company-specific modeling project– are designated here as "corporate terminology". This can relate to process objects (e. g. "purchase order", "invoice") or an activity (e. g. "repair", "check"). In most companies, various glossaries – normally partially redundant and inconsistent – define these technical terms with reference to standards (such as DIN). Technical terms are a subset of this corporate terminology. They are characterized by the fact that they are needed to describe details of the structure, i. e. they can potentially reference elements of a data model. The technical term model contributes to the definition of these terms, as well as to a systematic structure which includes explicit relationships between these terms. The distinction of technical terms by structure as well as by functions can also be found in the standardization work of the Open Applications Group (OAG).[60] The remaining terms are distinguished by other corporate terms (in particular names). In the following, only the technical terms are dealt with, not the other terms (refer to Figure 3.3).

Corporate terminology

Technical term

Function

Other terms

---

[58] See Ortner (1997), p. 32.
[59] See Spiegel (1993).
[60] See Ließmann, Engelhardt (1998).

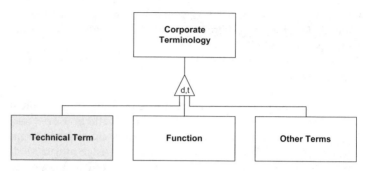

Fig. 3.3. Relation between corporate terminology and technical terms

3.5.2
Method for technical term modeling

Syntax: Description of notation

Technical term models are models with just one object type, i.e. technical term. Technical terms can designate quantities, features, and copies (instances). The critical factor is the creation of the right term and the definition of this term through a process of consensus. In addition, naming conventions[61] are of high relevance. Example: nouns have to be followed by adjectives – separated by a comma – (e.g. "facility, rented"). A mistake, which often occurs in practice, is the designation of technical terms using IT application systems (e. g. "SAP R/3 FI"), media, organizational units (e. g. "Z 6310") or states and / or generally imprecise designation of terms (e. g. "object data").

Attributes   The attributes of the technical terms are – in addition to the name (key attribute) – the definition with an indication of the source. This can be differentiated between a general dictionary such as a dictionary, and a company-specific definition. The attributes further includes examples, comments (e. g. about the degree of conformity with the term) and eventual short names. When defining technical terms, it has to be observed that the company-specific meaning is of primary interest, and not just the generally valid dictionary.

Relationship types   Technical terms are linked by pre-defined relationship types. In the following, the relevant relationship types[62] are outlined in technical term models (refer to Table 3.1, which contains samples from

---

[61]   See Chapter 3.4.3.
[62]   See also Spiegel (1993); Storey (1993).

the project). The opinion that technical term models should show only general relations but no concrete details[63] is not followed here since it is just the precise classification of these relations which essentially contributes to the understanding of technical terms.

A special feature of technical term models is the mapping of synonyms. This feature does not exist in data models. From two synonyms, one synonym has to be declared as the preferred term for further modeling. This term requires the attributes to be maintained, all other terms will reference this preferred term. Often, it is only possible to identify a synonym[64] by looking at the details of the characteristics of this technical term. In addition, a documentation of synonyms will avoid costly and often ineffective discussions about "uniform terms".

In addition, the difference between a "is part of" and "is content of" relationship has to be discussed. While "is content of" indicates that the sum of multiple elements in a "container" does not have any proprietary features which depend on the number of elements (a product catalog remains a product catalog even if 10 products out of 100 products are removed), the "is part of" relation indicates the fact that the total value only results from the interaction of single elements ("1+1=3-effect", e. g. a bill of material).[65]

Synonyms

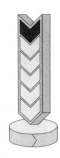

Table 3.1. Relationship types in the terminology model

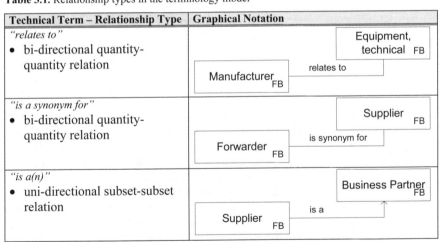

Technical Term – Relationship Type	Graphical Notation
"relates to" • bi-directional quantity-quantity relation	Manufacturer FB — relates to — Equipment, technical FB
"is a synonym for" • bi-directional quantity-quantity relation	Forwarder FB — is synonym for — Supplier FB
"is a(n)" • uni-directional subset-subset relation	Supplier FB — is a → Business Partner FB

---

[63] See Kirchmer (1998), p. 130.
[64] See Österle, Brenner (1986).
[65] See Storey (1993), pp. 463-465.

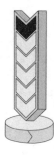

"is contents of" • uni-directional Subset – Container allocation	Product Catalog FB Product FB is contents of
"is a feature of" • uni-directional Feature – Quantity allocation	Customer FB Financial Standing FB is a feature of
"belongs to" • uni-directional Quantity – Quantity allocation	Department FB Employee FB belongs to
"is a copy of" • uni-directional Instance – Quantity allocation	Branch FB Chemistry FB is a copy of
"classifies" • uni-directional Feature (type) – Quantity allocation	Building FB Building Index FB classifies
"is part of" • uni-directional (Subset) Quantity – Integrity allocation	Floor FB Room FB is part of

Layout conventions

For the layout convention for designated relationship types, it is recommended that the source term be placed on the lower left side, and the target term – being the preferred term for synonyms – on the upper right side. In case of non-directed relationships, the master term according to the model context, is positioned in the upper right corner. The role of the relationship type should be arranged on the left side above the edge. This will create an easy to understand model (when reading from left to right) in the form of structured "technical term trees" (refer to Figures 3.4).

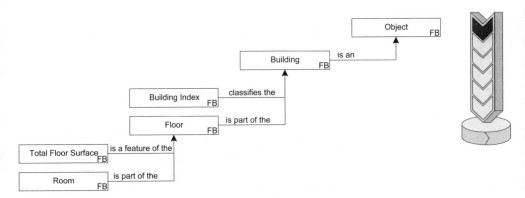

Fig. 3.4. Example of a technical term tree

Even though relation-related notes should be registered when creating the technical term model, the primary aim of technical term modeling is the creation of the term and not a complete data structure.

Syntax: Separation from other modeling techniques

Technical term terms are closely related to data and process models. A technical term can refer to a description of different elements in object and data models (e. g. object type, data cluster, entity type, reference type, attribute). In turn, an integrated and agreed technical term model represents an essential input for object and data modeling because it provides a high number of potentially relevant objects and related definitions. This way, the method experts can efficiently access the relevant terminology and their relationships to each other. But there are also technical terms which are not described in a data model (e. g. often "image"), because it does not seem relevant (at this time) to administer them in a structured format. In contrast, not every element of a data model necessarily corresponds to a technical term (e. g. the entity type "time"). Technical term models have to be assigned to the data view.[66] In summary, a technical term model can be separated from a data model as follows:

Data model

- Normally, a technical term model is clearly more detailed than a data model when compared with the number of object types, since it also describes objects which are of importance outside of application systems / database management systems.
- It uses predefined semantic relationship types (see Table 3.1).
- It shows the relations of synonyms.

---

[66] See Kirchmer (1998), p. 128.

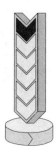

Process model

- Definitions in freely-formatted text dominate as attributes.
- It does not represent any cardinalities.

In particular, the consistency of process models can be considerably increased when an existing technical term model can be accessed.[67] Between process and technical term models, there are two relationships: First, technical terms can be used in process models as input and output data for the activities involved (refer to Figure 3.5). Second, when naming the functions it is recommended to define the process object as a technical term. Overall, a technical term model adds semantic information to the modeling conventions (Chapter 3.4). In addition, further object types (e. g. events in an event-driven process chain) can refer to technical terms.

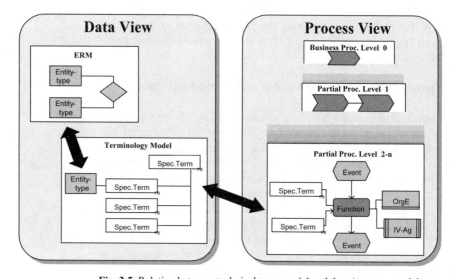

Fig. 3.5. Relation between technical term model and data / process models

Procedure model for technical term modeling

Selection of terms

When starting technical term modeling, all of a company's existing and / or relevant definitions of terms and glossaries have to be collected, verified, and consolidated. For this there are diverse sources available. The foundation is formed by the already existing, but mostly redundant and inconsistent definitions of terms of the company – the corporate term "legislative". Important informa-

---

[67] See Rosemann (1996a), p. 202.

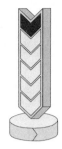

tion can be taken from product catalogs, product handling descriptions and the like. In addition, glossaries and terminologies from IT-applications already in use by a company or to be used in the future, as well as data models documenting the application systems, are considered in the technical term model.

From the stated terminologies, the technical terms are selected and assigned a unique definition, provided with the source, and integrated into the technical term model. It must be assured that every term is uniquely defined, that synonyms are recognized and eliminated, if possible, or else explicitly marked as synonyms, and that this terminology is constantly maintained. The more terms that are added, the more complicated and time-consuming their maintenance and integrity will be, i.e., the complexity of the technical term model increases. It is advantageous when the hierarchy of technical terms is object-oriented (refer to Figure 3.4) rather than function-oriented. This further simplifies, if applicable a later conversion in a data model which also focuses on object-orientation. The related control cycle is shown in Figure 3.6.

Unique definition

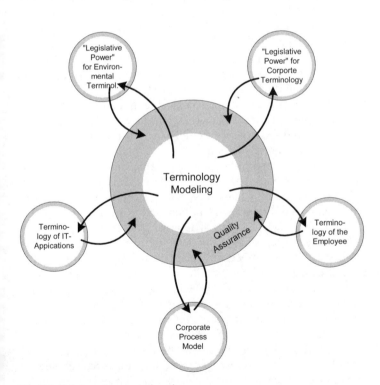

Fig. 3.6. Relations in a control cycle

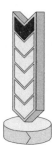

Quality check

After consolidation of the existing terminology, additional terms must then be identified and added to the technical term model. Prior to the use of a technical term model, the accuracy and suitability of the terms included must be checked. Users, who have a deeper understanding of the correct terms, perform the quality assurance of relevant terms. They check the terms for conformity with the corporate terminology, with the software-specific vocabulary, and with external definitions, such as legal terms. Technical terms need to be checked by experts for relationships, and whether or not the developed models match the modeling standard (Chapter 3.4.3).

Sources

One of DeTe Immobilien's first undertakings was to create the technical term model. This was performed even before the start of process modeling. First, all relevant sources were identified. These sources included numerous glossaries for different corporate areas, in particular definitions of terms from different projects, legal prescriptions (such as HOAI = payment regulation for architects and engineers), and various industrial standards. Besides this, the technical term model also included the results from data modeling within the scope of software development projects, definitions of terms from the company, and from Enterprise Systems to be implemented in the near future. In total, approx. 1,000 corporate terms were collected. Then, the relevant terms were selected, consolidated, and entered into the technical term model, together with the definitions of their related terms. The huge number of technical term terms, and the resulting complexity, led to high costs in the creation of this source solution.

Scope of modeling

Next, the as-is and to-be modeling terms were entered into the technical term model. This forced the terms to be defined immediately by users. Only in this way, could acceptable definitions be found and inconsistencies identified and corrected. Today, the technical term model consists of approx. 2,400 terms.

Due to the different degree of detail of the terms (e. g. "building index" vs. "facility agreement") and because of incompleteness as a result of the consequence of the prevailing focus on processes, the technical term model for DeTe Immobilien can only be used to a limited extent for software development projects. The existing technical term model can, however, provide a frame for terms, which can be applied to future IT-developments.

Suitability for process models

For third parties, the processes become understandable only when familiar terms are used as input and output for activities or as part of the designation of functions and events. This familiarity with terms decisively contributes to the acceptance of to-be processes by employees who are not involved in modeling.

The creation of a technical term model for DeTe Immobilien also meant a great deal of investment. The source solution needed approx. 5 man-months where parallel works could only be realized to a limited extent. Approx. 10 percent of the time of the to-be processes modeling was spent on technical term modeling. During this time, one additional employee was needed to update the terminology and to maintain the relationships of the terms.

Investment

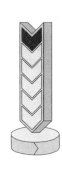

3.5.3
Distribution and maintenance

The technical term model of DeTe Immobilien, maintained within the ARIS-Toolset, is communicated in a graphic format and in two other formats in order to be distributed to the entire enterprise. First, a report was created in a structured ASCII file with the terms sorted in alphabetic order and distributed in limited quantity.

Report

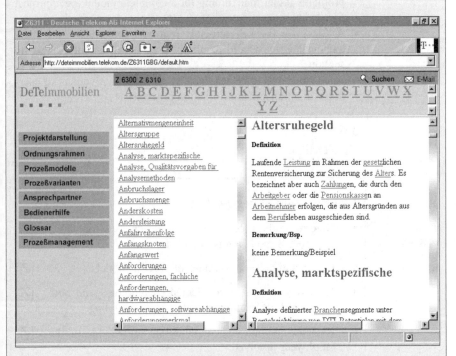

Fig. 3.7. Extract from the Web-based technical term model

The further distribution of technical terms via the Intranet enabled a much more up-to-date set of terms to be available to more employees. The related ASCII text was imported into a database and

Intranet

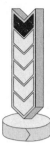

then converted into HTML format, together with related hyper-links, which were recognized by the arrangement of terms and identified by the repetition of terms. This way, all employees could access the terminology via a PC. A simple feedback mechanism allowed the consolidation of valuable input from the organizational members. An example of this Web-technical term model is given in Figure 3.7. Though this is only in German, it will give an idea of the structure and user interface. This technical term model is updated at regular intervals and points to other special terms (e. g. DIN). An intuitive WWW-based representation of a terminology system with comprehensive retrieval and navigation options has also been developed within the GIPP-Project[68].

*Recommend-
ations for
design*

The following recommendations can be given for the design of technical term models.

- Technical term modeling should primarily focus on the actual problems rather than aiming for a complete model, i.e. a project focused approach instead of a corporate-wide approach.
- A focus on the key terms is required. Otherwise, too many technical terms would result for a relatively small real world section.
- Acceptance has to be achieved by an early involvement of users in the creation and consolidation of terms.

3.6
Checklist

Identification of perspectives

*What to
observe!*

- Determine the potentially relevant purposes for which ☑
  the process modeling is intended.

- Create and present examples of process models. ☑

- Hold discussions with selected stakeholders and note ☑
  their requirements in terms of process modeling.

- Rate the methodological modeling knowledge of the ☑
  employees involved.

---

[68] See Hagemeyer, Rolles (1997).

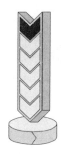

- Consolidate the requirements of the process model. Define which perspectives have to be differentiated and how. ☑

Modeling technique and modeling tool

- Derive the requirements of the modeling technique and of the modeling tool from the requirements of the stakeholders. ☑

- Note other areas within the organization which are interested in process modeling and a modeling tool (e.g. documentation of IT-infrastructure, data modeling). ☑

- Check which modeling tools are already used in the organization. ☑

- Select the modeling technique and a modeling tool which supports the requirements. ☑

Modeling conventions

- Rate the importance of correctness, relevance, economic efficiency, clarity, comparability, and systematic design for your purposes. ☑

- Consolidate the already existing conventions in your company as well as any conventions contained in the modeling tool. ☑

- Create your own individual conventions, by defining the use of the relevant modeling techniques, object types relationship types and attributes. ☑

- Configure the modeling tool based on the agreed modeling conventions. ☑

- Train the employees involved in modeling, and make sure that the modeling conventions are made available for reference and can be easily accessed. ☑

- Integrate new important conventions into the agreed modeling conventions as they emerge through the modeling process, and distribute these changes. ☑

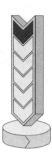

Technical term modeling

- Collect the existing glossaries, relevant laws and similar documents. ☑

- Record the definitions of the technical terms in agreement with the users and together with the relationship of these technical terms to each other. ☑

- Consolidate the technical terms and communicate them in a suitable form (preferably via Intranet). ☑

- Reference the technical terms during process modeling or complete the technical term models, which include the technical terms, in the course of process modeling. ☑

From Strategy to the Business Process Framework

Jörg Becker, Volker Meise

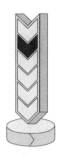

4.1
Organization design in strategic context

Designing the structure of an enterprise is a task which is of strategic dimension because of its long-term and fundamental effects. Therefore, considerations about the basic design of structures and systems are focal points in strategic management.[69] Strategic management is intended to derive planning, control, and coordination of business development with an holistic view.[70] It is not the reactive adaptation to environmental changes that is regarded as being the contents of business development, but the procactive design of external environmental relations and the internal configuration of the enterprise (organizational structure and corporate culture).

Pragmatically, the task of strategic management is described as searching, building, and upgrading, as well as maintaining strategic potentials for success.[71] Not infrequently, the related measures constitute the total value chain of the company. The organizational structure serves as infrastructure which influences the creative development and successful conversion of new ideas with a long-lasting effect. Organizational innovations are high-ranked strategic resources to gain competitive advantages.[72]

Strategic management

Organization in the sense of the organizational term "structured design,"[73] goes far beyond the "coordinatation of actions which regulates the interaction of people and people, people and object, and object and object, in order to achieve the defined objective". A

Organization

---

[69] See Bleicher (1996), p. 75.
[70] See Welge, Al-Laham (1992), pp. 2355-2357.
[71] See Gälweiler (1990).
[72] See Frese, v. Werder (1994), p. 4.
[73] Kosiol (1976), p. 20.

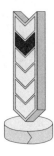

Structure follows strategy

Change of organization

structured design is not only a prerequisite for the functioning of this interaction between all existing resources in a company, but also plays a decisive role in competition. This requires the organizational structure to orient closely to the strategy in order to support strategic navigation.

CHANDLER's strategy was to define long-term goals, to assign existing and expected resources required for the achievement of the defined goals, and to decide on goal-adequate actions. This strategy, developed in 1962, as "structure follows strategy,"[74] is still valid today. The analysis of organizational changes in the German economy[75] has shown that it is extremely important to watch the competitive situation thoroughly while restructuring and changing in order to recognize strategic priorities which have to be supported by organizational structure. The consequential derivation of the structure from the strategy of a company is considered ideal because every formulated strategy in a company needs to be adapted to conform with both the organizational structure and the business processes in order to foster the new orientation[76]. In enterprises where the structural prerequisites are missing, the implementation of new strategies usually fails.[77]

Figure 4.1 depicts as an example the evolution of the role of production as a result of changed market requirements. The strategy has to be adapted to the related evolution of the competitive situation in order to cope with the new requirements and to secure the competitiveness of the company. Therefore, the organizational structure has to be corrected as well.[78] An organization that desires to fulfill its customers' needs in regard to fast processing and delivery of orders, and that wants to achieve this by "streamlining" its processes, will conceivably have problems with the creative process in designing innovative, unique products.

---

[74] Chandler (1962).
[75] See Arbeitskreis 'Organisation' of the Schmalenbach-Gesellschaft (1996), p. 641 und pp. 643-644.
[76] See Andrews (1987), pp. 20-21 quoted in Thiele (1997), pp. 18-19.
[77] See PoMez (1992), pp. 166-168.
[78] See Arbeitskreis 'Organisation' of the Schmalenbach-Gesellschaft (1996), p. 44.

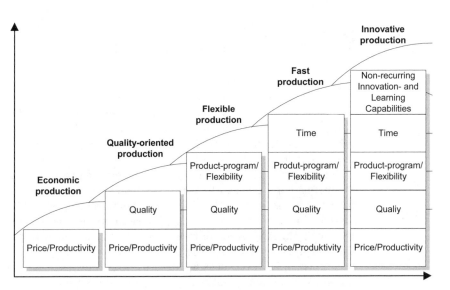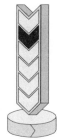

Fig. 4.1. Evolution of production as a result of changed market conditions (Source: Zahn (1994), p. 245).

The adaptation of strategy and structure, in particular their timing and quality, widely impacts the success of a company.[79] Therefore, an intensive dealing with fundamental strategic alternatives and their subsequent effects is mandatory.

Importance of strategy

While in the 80's a strictly market-oriented external strategic view prevailed (market-based view), this view has changed in recent years to a more resource-based internal view (resource-based view). Often, these two basically different views are regarded as being bilaterally exclusive. After critical consideration, however, this opinion cannot be supported. A company management which exclusively orients toward core competencies is as one-sided as a company management which follows a strict market orientation. The aim should be to combine both views, i.e., the market-based and the resource-based view, to one total view. These considerations have to be preceded by decisions on relevant markets or market segments, which will then lead to the creation of strategic business fields and units.

Market-based and resource-based view

---

[79] See Müller-Stewens (1992), p. 2345.

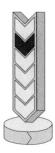

4.2
Strategic business fields and units

Before an enterprise can make decisions on its structure, it has to determine in which markets or market segments it wants to become active. The items to be defined are strategic business fields (SBF) whose organizational equivalents are found in strategic business units (SBU). Only after this can individual strategies be defined within the SBU.

Definition of strategic business fields

The definition of strategic business fields mainly orients toward the internal homogeneity of the business field.[80] This is required in order to define dedicated market exploration strategies and to measure the success of these strategies based on the related business field. The subdivision of business fields can be done with different degrees of detailing. The simplest division is by products. This division, however, does not meet the requirements of a market-oriented corporate strategy.[81] An expansion by market dimension leads to a definition of the SBF as a product-market area which enables a differentiated view.[82] A further differentiation creates a 3-dimensional reference frame between user group, functionality, and technology.[83] A further refinement step can be achieved by taking the restricted regional areas as selection criteria.[84]

Homogeneity of SBF

It must be noted that the homogeneity of strategic business fields tends to increase with the increased degree of detailing – while the independency of the success factors decreases to the same extent. With other words, the smaller the strategic business field is, the higher is the propability that it has to share resources with other SBFs.

A generally valid solution for business field definitions does not exist. Depending on market size and market features, different degrees of detailing can be executed. As heuristics, an iterative procedure is recommended to test the requirements of independence. In case of doubt, a strategic business field should be extended in order to avoid the risk of loosing chances in single market segments.

Definition of SBU

The definition of strategic business fields is followed by the definition of strategic business units. They form clearly defined units of the company and represent the organizational conversion

---

[80] See Hinterhuber (1992), pp. 72-73.
[81] See Meffert (1998), p. 226.
[82] See Hinterhuber (1992), pp. 107-108.
[83] See Abell (1980).
[84] See Meffert (1998), p. 228.

of the specified business fields, including an independent definition of strategy. From the customer's point of view, and in the sense of a uniform market presence, it is reasonable to implement one strategic business unit in one business field. However, it can also be reasonable in certain market-customer combinations that the business fields of two or more business units overlap.[85]

The positioning of different business units in a company can be performed using the portfolio-analysis methods – the most popular market share / market growth portfolio of the Boston Consulting Group[86] can be stated here as an example.

DeTe Immobilien identified three main business fields in their strategic analysis. The most important business field with the highest turnover is facility management for Deutsche Telekom AG. Besides this main business area, the intention was to expand the facility management to customers in the free market as a secondary supporting system. The products and services did not differ, in essence, from the products and services provided to the parent company. They differed only in the success factors because the services were partially calculated with the parent company on a lump sum basis. The third strategic business field is the development of large real estate projects for external investors. Here, processes are concerned which range from the creation of first site analysis up to the delivery of ready-made and rented building complexes. The customers mainly consisted of institutional investors and, therefore, differ primarily from those of the first two business fields.

Business fields of DeTe Immobilien

4.3
Market-based view of strategy

Both the market- and the resource-based view try to explain the prerequisites and conditions to achieve a long-lasting return on investment for the company. From the market-based point of view, advantages in competition arise primarily from the optimized positioning of the company in an attractive special line or in a group of strategic importance. Market entry barriers and / or mobility blockers should prevent competitors from penetrating the

Positioning of company

---

[85] This is the case with digital photo cameras. The products address the PC user who wants to use the photos in digital form, e.g. for the Internet, and the products also address the ‚normal' user who values their ease of use.

[86] See Hedley (1977), pp. 10-12.

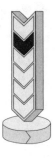

Low cost leadership or differentiation

market, and this should secure the scope and durability of strategically long-lasting profits.[87]

The question is which strategies can be used to create such barriers for potential competitors. PORTER recognizes only two basic types of competitive advantages which a company can make use of: low costs or differentiation.[88] If these competitive advantages are combined with the activity field of the company, then strategic leadership in low-cost or product differentiation will result in an above-average turnover in a special trade line (refer to Figure 4.2).

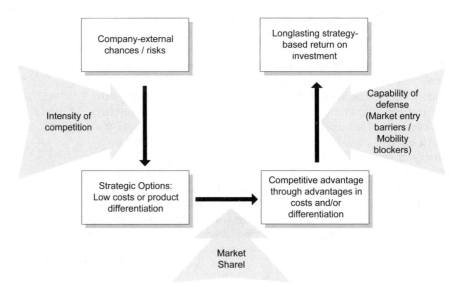

Fig. 4.2. Market-based view of strategy (as per Barney (1991), p. 101)

4.3.1
Low-cost leadership

Detection of possible cost savings

The strategy of low-cost leadership[89] is based on a learning curve concept which says as a rule of thump that the doubling of the cumulated production quantity results in average cost saving of 20 to 30 percent. This strategy aims at the company becoming the sole most cost-effective manufacturer within a special line (branch) due to comprehensive price advantages. The approaches

---

[87]   See Gaitanides, Sjurts (1995), p. 62.
[88]   See Porter (1992); Porter (1989).
[89]   See Porter (1992), pp. 63-63; Porter (1989), pp. 31-33 for the following.

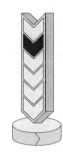

to realize this cost advantage include: upgrading of production equipment to efficient sizes; full utilization of known cost reduction potentials; stricter control of variable costs and overhead costs; avoidance of marginal customers; cost minimization in the indirect areas. The detection and utilization of all sources for cost advantages have to be given highest priority.

When a company has achieved low-cost leadership, the company will experience an above-average outcome in its special trade if it succeeds in establishing prices on the same level or in proximity to the average price level of the branch. The superior cost structure ensures a higher contribution to cost coverage than that of competition.

Normally, the factors which have led to cost advantages create considerable market entry barriers against potential competitors. The cost advantages which are based on the reputation of the company, or the experiences in cost-minimized production of products, cannot be immediately realized by new competitors just entering the market.

Market entry barriers against competitors

4.3.2
Differentiation

The strategy of differentiation from competition is to achieve a unique position. The company selects one or more features which representative customer groups deem to be essential. Then, attempts are made to achieve this unique position by meeting the exact requirements of these representative customer groups. Such a unique position cannot be substituted by products from competition. It binds the customers to their own brand and makes them less sensitive to prices. This, in turn, enables higher cost-covering margins unless the company disregards its price position and achieves an equal or approximate price parity without loss of differentiation.

Aiming at a unique position

The differentiation strategy requires that the company succeeds in being unique, or is regarded as being unique by at least one differentiating feature. Otherwise, it will not be possible to achieve higher prices – consequently, income will be below-average in comparison to that of the low-cost leaders because of the diminished cost position. In contrast to low-cost leadership, there may be various differentiation strategies in a branch when multiple features exist which are recognized by the customers. In many markets, a differentiation stragegy can be observed which accentuates products by focusing on special design and superior materials. Here, the combination of extraordinary design and selected materials makes the difference to strategies which intend to differentiate

Various differentiation strategies

their low-cost leadership by exclusively concentrating on the design.

4.3.3
The value chain

In order to view the competitive advantages within a company, PORTER creates a working model of the value chain by arranging and portraying all "activities" necessary to design, manufacture, sell, deliver and support a product[90] (refer to Figure 4.3). He bases this on a classical division of functions in the business areas of the company which are supported by similar classical cross sections.

Supporting Tasks	Corporation Infrastructure					Profit Margin
	Personnel Management					
	Technological Development					
	Procurement					
	Reception Logistics	Operations	Marketing & Sales	Delivery Logistics	Customer Services	Profit Margin
	Primary Activities					

Fig. 4.3. Model of a value chain (source: Porter (1989), p. 62).

Value chain as structural frame

PORTER analyzes the individual areas of the value chain by their potential contribution to low-cost leadership or differentiation strategy. This information, however, is difficult to convert to concrete design criteria to structure the value chain elements in the specific situation of the company. The value chain itself represents a certain business process framework already – in a very general form – as used for process-oriented corporate structuring.

4.4
Resource-based view of strategy

While the market-based view determines corporate strategy directly, in terms of the line of business and the behavior of

---

[90] Porter (1989), p. 63.

competition, the resource-based view is based on the internal view. *Internal view* This view analyzes the strengths and weaknesses of the company *of the* and derives development direction and strategy from them. The *company* optimized utilization and expansion of these resources secure competitive advantages over other market participants who do not possess appropriate resources or the possibilities to utilize these as differentiation features.

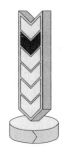

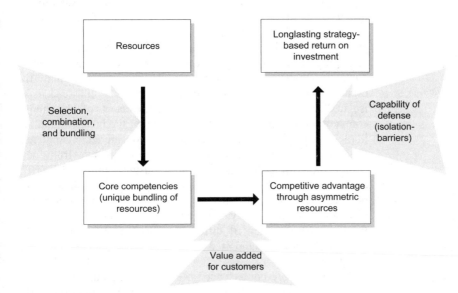

Fig. 4.4. Resource-based view of strategy (as per Barney (1991), p. 101)

The focus on areas of one's own strength and capability makes *Focus on own* sense not only from a competitive point of view, but also in regard *strengths* to transaction costs[91]. When the corporate strategy is aimed to-wards an increased diversification in order to use technology-based cost digression effects – which result from the increasing size of the company – then there is the risk that the increasing costs of communication, coordination, and decision-finding over-compensate the savings which result from the scaling effect of the company size. Successful companies, therefore, often focus their efforts on a few business fields only.[92]

The concept of core competencies deals with the identification, *Competence* evaluation, and extension of these core fields and capabilities. *as capability*

---

[91] Transaction costs are the costs associated with the market exploration, the design of the contract, the completion and the control of the provisions.
[92] See Albach (1988), p. 81.

Rather than interpreting the terms "right" or "entitled to act," which are mainly used in organizational documents, the term "competence" is interpreted here with "capability".

4.4.1
Core competencies

Criticism of SBU orientation

Almost no strategy concept has been given as much attention in the 90's as that of the core competency approach of C. K. PRAHALAD and GARY HAMEL,[93] published in 1990. Both authors criticize the hitherto prevailing strategy orientation of using strategic business units (SBUs) and support a competence-oriented company management. They are of the opinion that the organization which orients itself toward SBUs would lead to an isolated consideration of just one single facet of the global competition, i.e., the competitive products of the business unit are implemented at the time when this is considered. Completely forgotten were the opportunities that would result from new combinations of capabilities of single business units. The SBUs are trapped in a corset which enables only near developments to be realized, i.e., marginal extension of product line or geographical expansion. Radical innovative and unique products which result from the combination of different technologies would only be successful when supported by the management of core competencies.

The term "core competence"

The term "core competence" needs to be explained. Every enterprise possesses a number of technologies and employee capabilities to develop, manufacture, and sell products. The bundling of these capabilities and technologies leads to single competencies. These single competencies are not yet core competencies. Only after single competencies have been integrated in one, new, overall and hard-to-emulate capability will this lead to a true core competence. A core competence can hardly exist for one person or smaller groups, but rather represents the cumulated know-how of overlapping individual capabilities or organizational units.

Linking of core competencies

The linking of different core competencies in a company enables it to develop core products, forming the basis for final products in the individual business units. For example, the production know-how for the manufacture of LCD screens can also be used in the PC area, as well as for video cameras, in TV, or for digital photo cameras. The core competencies can be compared to the roots of a tree (refer to Figure 4.5).

---

[93] See Prahalad, Hamel (1990).

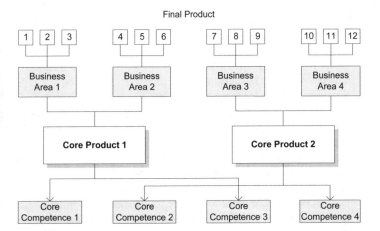

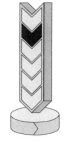

Fig. 4.5. Competencies: The roots of competitiveness (as per Prahalad, Hamel (1990), p. 81).

The difference between a particular capability and a core competence is hard to define. In principle, a core competence has to fulfill three criteria:

- Creation of value for customers.

  A core competence has to enable a company to create basic value for the customers – where the customer-view decides whether or not a core value exists. It does not play any role in whether or not the customer understands which competence is bringing him this value. An example for the core value is the know-how of a car manufacturer in the construction of an engine. The customers decide on a brand (in this case a car) because of the economic performance and power of the engine – and the extent of the dealer network. The dealer network also requires a competence in establishment and service, but this is regarded as "normal" and will not be the main criteria for a decision in favor of this type of car. An exception in the criteria of value for customers is the capability to manufacture a product efficiently. Even if the cost advantage over the competitor is not passed on to the customer and, therefore, the customer does not extract any direct value from this competence, it is nevertheless regarded as core competence. *Value for customers*

- Enabling a differentiation of competitors.

  A core competence must be unique within the competitive environment. This does not mean, however, that the related company is necessarily the only company which possesses this competence. Instead it means that the competence, in comparison with that of the competition, is dominant. Therefore, de- *Differentiation of competitors*

Portability to other business fields

pending on the market segment in which it is used, the core competence must be assessed. A non-core competence in a major business field can very well become a core competence when transferred into a new or regained market. This corresponds to the next criterion:

- Securing upgradability and portability.

Mostly, the core competence is closely connected to existing products. It must be possible, however, to detach this competence and to transfer it to other business fields. If this strategic relevance is given on multiple markets, then synergy effects can be realized through the utilization of core competencies.[94] When indicating core competencies, an abstract view is required in order to be able to separate the competencies from the strong focus on products and to reveal possible transfers options.

Long-term importance

In addition to these three criteria, there is still one criteria left to be mentioned and which must not be omitted in the strategic dimension of identification and extension of core competencies. The core competence must be of long-term importance and must not be able to be imitated by competition in the short run. The durability has to be seen in relation to the innovation cycle in the branch.[95]

What are core competencies; what is a particular capability? A plant, a sales channel, a brand, or a patent is not a core competence. Nor do core competencies appear on the balance sheet. But the capability to manage plants (Toyota's lean manufacturing), sales channels (Wal-Mart's Logistics), brands (Coca-Cola's) or proprietor potential (Motorola's capability to protect and use their patent-portfolio) fulfils the criteria of value for the customer, of uniqueness, and of portability.[96]

Equality of views

An enterprise should not only be viewed as a portfolio of products, but also as a portfolio of core competencies. PRAHALAD and HAMEL underline the equality of both views – in contrast to the many succeeding defenders and critics of the alleged completely revolutionary management approach. In spite of all this, they leave no doubt about the strategic importance of their consideration: "Tomorrow's growth depends on today's competence-building"[97].

---

[94]　See Stalk, Evans, Shulman (1992), pp. 65-67.
[95]　See Thiele (1997), p. 76.
[96]　See Prahalad, Hamel (1994), pp. 207-208.
[97]　Prahalad, Hamel (1994), p. 222.

4.4.2
Management of core competencies

The decision to manage a company by core competencies must also be followed by the identification of existing core competencies and their management. It is recommended to do this in four steps. In the first step, the core competencies of the company have to be identified in order to back up and extend them. Since core competencies can also become vulnerable in the course of time, the second step should include the creation of new competencies which could be extended later. The last two steps will not be explained in detail because they are not of essential importance for the derivation of the business process framework for process modeling.

The 4-step approach

Step 1: Identification of core competencies[98]

A strategy towards core competencies cannot be successful when all participants do not have the same concept of what core competencies are and which effects they have. Therefore, it is absolutely mandatory to clearly and uniquely define the core competencies of the company. This definition has to be understood and agreed upon by everyone. A mistake to be avoided is the use of the smallest common denominator for the definition of core competencies in order to achieve the widest possible acceptance. Many employees will have an approximate idea of what a company "can do very well", but they have problems building a relationship between the bundles of unique capabilities and the success of the final products which result from these capabilities.

Unique Definition of core competencies

This leads to the fact that equipment, infrastructures, or patents are misunderstood as core competencies, and when in doubt, any person involved believes that his task is an indispensable feature of a core competence. These beliefs cause a long list of activities to be created which are claimed to be core competencies. This list must, therefore, be reduced to a limited number of competencies that can be handled, unless this has already been done formerly by strict disciplinary tests with the three above-mentioned criteria. The aggregation should be selected in such a way that 5 to 15 core competencies remain valid rather than 40 to 50 core competencies.[99] In particular, it is essential to take the market-based view as the central criterion, i.e. the perception of customers.

Reduction to 5 to 15 core competencies

---

[98] See Prahalad, Hamel (1994), pp. 224-226
[99] See Prahalad, Hamel (1994), p. 203.

Step 2: Create core competence portfolio

After the definition and evaluation of the existing core competencies of a company, the next thing to do is to create the portfolio from existing and future core competencies and to thus set directives for the development of the company. A classical four-fields-portfolio can be set between the core competencies and the market, divided by existing and new competencies and / or market segments (refer to Figure 4.6).

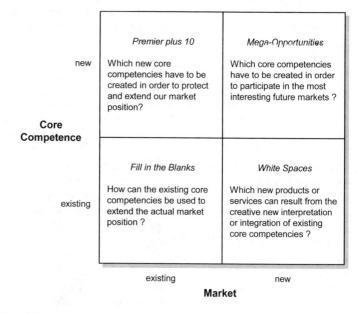

Fig. 4.6. Core competence portfolio (Source: Prahalad, Hamel (1994), p. 227)

The analysis of the portfolios leads to strategies of individual core competencies: maintaining, extending, or a more intensive use. Precisely these directives characterize the organizational structure of business fields within a company that deal with these individual core competencies.

Core competencies of DeTe Immobilien

The analysis of the core competencies of DeTe Immobilien showed competencies in the setup and maintenance of technical equipment of buildings, in the planning services for facility management, and in the development of real estate. The integration of these competencies and the resulting option to offer overall facility

management – from the planning of the building, through setup and equipment, up to the operation of real estate – now represents the competitive advantage of DeTe Immobilien.

4.5
Combination of views

The core competence approach is based on the resource-based view of competitive advantage, and attempts to explain the differences between companies (and their different degrees of success) by their different, unique resources. As explained above, this approach is not new in principle. The enthusiasm with which the article of PRAHALAD and HAMEL had been accepted, and the following wave of publications, was amazing. In particular, the publications which euphorically discussed a new management paradigm often overlooked the fine details in the statements of PRAHALAD and HAMEL. Consequently, the responses were of similar black-and-white coloring.[100]

Meanwhile, the core compentence approach is judged to be an important contributor to the explanation of potential success. This approach is used in addition to the existing market-based view. However, management purely based on core competencies is as one-sided as a complete market-based view. On the one hand, core competencies can become obsolete more quickly than anticipated, either by legal prescriptions or by quick-learning competitors. On the other hand, a focus on a few markets can lead to overlooking the possibilty of transferability of the company's own capabilities and therefore, a more balanced use of resources with additional chances for profit.

Balance of market view and core competence view

Core competencies are not the only criterion for the organization of a company, but complete the strategic business units. These are described in Chapter 4.2, and primarily result from the market-based view. The core competencies and the resulting core products are assigned to strategic business units from the company-internal view, based on an analysis of its own resources. The SBUs relate to the core competence-oriented structure of business areas. On the one hand, the evolution of existing products and the development of future products is initiated by the SBU and by the demand of the customers; they define the requirements of the products. On the other hand, there is the core-competence oriented view. It ensures an optimized allocation of resources within the fields of core competencies and provides different core competencies with the possi-

Relation of SBU to core competencies

---

[100] For this phenomenon, refer to Kieser (1996).

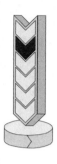

SBU accesses the resources

bility of offering the SBUs new and innovative products; in particular, by combining core competencies.

In addition to the high innovative power, the core competence orientation leads to an increase in the use of resources when multiple SBUs access the resources of one core competence. Within the core competence fields, the structural criteria are most often designed for an optimized utilization of resources. An exception is when the unique mastering of the process itself represents a core competence. In this case, priority is given to the optimization of the process flow.

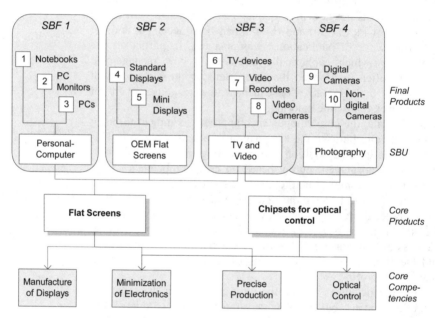

Fig. 4.7. Example of a corporate structure by strategic business units and core competencies

The combination of market-based and resource-based views enables the requirements of the customer to be met, and at the same time avoids missing the chances offered by the innovative combinations of branch-overlapping core competencies. Figure 4.7 shows examples of how the structure of a company may look.

4.6
From structure to core processes

The structure of a company area is determined by different influence factors. On the one hand, the core competence and its orientation require a certain organizational structure (the core competence can belong to this area in part or in full). On the other hand, the structure is also influenced by the status of the life cycle of an organizational unit. When a company unit is developing, it will need the same structures as R&D- and core competence-oriented areas which support the learning process from outside and the management know-how of internal resources – a "learning organization". In the life cycle to come, the focus is more on the efficient utilization of resources or on unique process mastering (refer to Figure 4.8). All these different objectives need tailored organizational structures. "A world-class competence must steer the power structure in a company".[101]

Degree of structuring

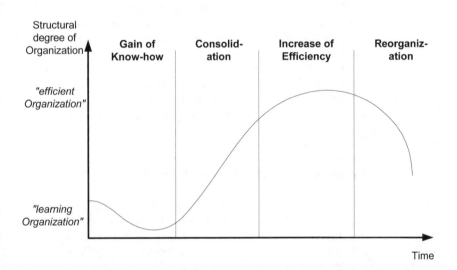

Fig. 4.8. The Lifecycle of Organizational Units

Consequently, clear concepts have to exist prior to restructuring as to which organizational concepts can be reasonably transferred to its own enterprise or to which units.[102] A adoption of organization-

From strategy to structure

---

[101] Coyne, Hall, Gorman Clifford (1997), p. 48.
[102] See Kieser (1997), pp. 99-101.

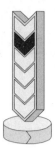

al concepts off-the-shelf is risky and will seldom provide success.[103] What is required is a stringent derivation from strategy to structure – operational changes are legitimized only by the performance request of the strategy.[104]

4.6.1
Use of efficiency criteria

Evaluating alternatives

In order to avoid that restructuring becomes a journey "where nobody knows the destination or travel time" [105], criteria have to be developed to decide on and measure the changes. A suitable evaluation scheme would be the comparison of efficiencies.

Coordination- and motivation efficiencies

On a high level, efficiency criteria can be distinguished by co-ordination- and motivation efficiencies.[106] The coordination view deals with the optimization of autonomy and reconciliation costs resulting from work sharing. It can be divided into four criteria: market efficiency deals with the use of potentials on external procurement and sales markets; resource efficiency deals with the exploitation of corporate potenial factors; process efficiency deals with optimized design of business processes; and delegation efficiency deals with the delegation of decisions between hierarchy levels.[107]

Subdivision of coordination efficiency

Since predicted behavior and actual response of the involved individuals do not always match, additional criteria have to be observed that influence the motivation of the employees in the sense of a coordination-adequate behavior. It is the motivation view that deals with this topic.

4.6.2
Example: Business Process Reengineering

BPR as solution to all problems?

The decisive point is that efficiency criteria cannot always be harmonized. This can be demonstrated very well with the success and failure of reengineering measures. Business Process Reengineering (BPR)[108] is strongly process-oriented, which means that process efficiency has priority over all other criteria. The use of BPR-approaches for the reorganization of business units – whose

---

[103] See Arbeitskreis 'Organisation' of the Schmalenbach-Gesellschaft (1996), p. 622.
[104] See Hagel III (1994), p. 105.
[105] Champy (1994), p. 88.
[106] See Frese (1995), pp. 269-271.
[107] A detailed description of efficiency criteria is given in Chapter 7.2.1.
[108] See Hammer (1993).

task is the fast and efficient processing of business processes – fulfills the often-stated radical improvement with cost savings.[109]

The reengineering debate, however, has gone its own way for a long time already, and Business Process Reengineering and its derivatives are praised as being new organizational paradigms which overcome all the problems of a company. An analysis of whether or not the radical focussing on processes is reasonable for all areas is performed rarely. Therefore, it is no wonder that BPR-projects fail on those occasions when the success of the company unit is based on resource efficiency rather than on process efficiency, and on the optimized utilization of machines rather than on shorter response times. The improvements through the best possible fulfillment of process efficiency is over-compensated by the disadvantages of the other efficiency types. The harmonization of goals as proclamed by the BPR method cannot be maintained.[110]

Setting of efficiency goals

However, we have to abstain from favoring a solution where the smallest common denominator is searched for among the criteria – this is the traditional approach and hinders the chance for comprehensive improvements. The company must be aware of which efficiency goals in which areas have to be prioritized . It is important to analyse which combination of efficiency criteria secures long-lasting corporate success and whether or not different combinations have to be applied to different corporate sections. When sorting areas by core competencies it can be assured that the efficiency goals within an area are homogeneous. But when considering traditional areas such as R&D or Logistics, it is also obvious that internal efficiency goals clearly differ from those of other areas.[111]

4.6.3
Derivation of core processes

The definition of efficiency goals for superior areas is followed by a similar structuring of the subordinate units. Even within an area which, in total, aims at the greatest possible resource efficiency, it may happen that strict process orientation must be optimized, and vice versa.

First, it is important to arrange the core processes on the upper level. The core processes result from the described strategic company analysis and the derived tasks. They represent the main tasks of the company and, therefore, differ from the support or enabling processes which have a supportive role. The core processes must be described in more detail. While the modeling of the total enter-

Core- and support processes

---

[109] See e.g. Champy (1994).
[110] See Theuvsen (1997), p. 122.
[111] See Hagel III (1994), p. 100.

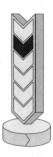

Core processes have steering functions

prise can depict the strategic business unit as a process, e.g. "product sales on market X", the modeling of the business unit itself includes the detailed subdivision of the process.

On a total company level, and after structuring, the core processes can be divided into strategic business units and fields of core competence. This way, structurally equal processes of different strategic business units, such as sales, can be portrayed as one core process. The splitting into different variants is done in the phases of to-be modeling or when deriving the new organizational structure.

A process-focused orientation of the company is often equated with the demand that the core processes go "from customer to customer (C2C)", so-called wall-to-wall processes. A company whose strength is the customized production and delivery of its products can define a core process "production and sales". However, if the company's own products and the purchase of foreign products are equal in weight, then it makes more sense to have the two core processes "production" and "purchasing," access the core process "sales". Even if this division resembles an orientation toward traditional business functions, the assignment of a person to be responsible for a particular process, for instance, can be determined based on the process-oriented arrangement and the coordinated interfaces of the core processes. In this way, continuous process thinking can be realized.

Core Processes at DeTe Immobilien

The core processes of DeTe Immobilien have been developed by a strategy team. The processes are subdivided into market-oriented and production-oriented components (Figure 4.9).

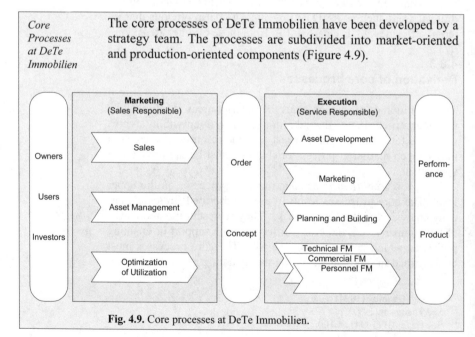

Fig. 4.9. Core processes at DeTe Immobilien.

After this division, the sales department should form the only interface to the customers, in the sense of "one face to the customer," and give order to the fulfillment processes. These processes were divided into development of single assets; marketing in the sense of broker activities; planning and building using architectural and engineering services; and the execution of technical, commercial and personnel facility management. The processes of asset management and optimization of resources as well as the sales process have responsibility for revenue. They execute the value analysis of the portfolio of facilities and are responsible for the surface- and cost optimization.

One face to the customer

When discussing the defined core processes, problems arose in the delimitation and reconciliation of the single processes. In this first draft, the identification of the core processes referred too much to the existing organizational structure. In order to bundle the competencies further, and to optimize the cooperation of the different processes, it was decided to summarize the optimization of utilization and parts of the process planning and building because of their similar competencies. The result was the core process "planning". The "milestone" within this process was the conclusion of the contract, followed by a consultation with the customer, and a subsequent detailed planning. Another summary of core processes resulted from marketing. Marketing has already been regarded in the first draft as closely connected with the sales process and was, therefore, assigned to the sales process in order to minimize the interfaces.

Problems with core processes

4.7
Business process framework as guiding principle

The findings from the analysis of the strategic orientation of the company and their effects on the organizational structure should be summarized to serve as a guiding principle for the processes to follow. The most suitable way of doing so is to use a business process framework. A business process framework divides the structures of the company on an abstract level by a selected organizational paradigm and clarifies the relationships between the individual parts of this framework. The Retail-H[112] is such an example.

Abstraction and structuring

The Retail-H has been developed to divide and arrange reference function-, data- and process models for retailing companies. In the Retail-H-Model, the major tasks of retailing are

Retail-H-Model

---

[112] See Becker, Schütte (1996).

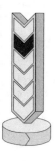

listed. The sections procurement and distribution are linked by the bridging function of the store (warehouse). Procurement includes the timely sequence of subtasks such as contracting, order management, goods receipt, invoice auditing, and accounts payable. In an analogue structure, the distribution process consists of marketing, sales, goods issue, invoicing, and accounts receivable. The left side contains the tasks with reference to the supplier, the right side contains the tasks with reference to the customer. The enabling processes general accounting, asset management, cost accounting and human resource management form the foundation of this framework. The "roof" of this model is provided by the strategic processes controlling, Executive Information Systems and strategic planning. The basic processes of retailing and the interaction of the functional areas can be displayed by inserting and removing tasks (refer to Figure 4.10).

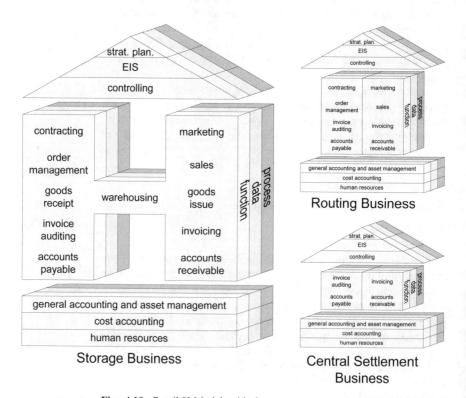

Fig. 4.10. Retail-H-Model with its occurrences as storage-, routing- and central settlement service (Source: Becker, Schütte (1996), p. 11 and p. 420)

A business process framework systemizes the processes on the top level – the core processes – in the sequence they are executed. For more detailed process modeling, the business process framework *Arrangement* provides an important guideance when analyzing the arrangement *of the core* of the individual functions and processes. It also helps to identify *processes* and structure the interfaces of the subordinate processes.

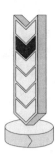

The business process framework is typically further developed and revised in the detailed process modeling phase when further information about the structure of the business processes are available. This causes a dilemma in the design of the business process framework. On the one hand, a business process framework should already exist when modeling starts in order to serve as a guiding principle and to arrange the models to be created. On the other *Development* hand, the blueprint for the new organization – both processes and *of business* structure – only results from the new processes being modeled. *process* While it is important to adapt the business process framework ac- *frameworks* cording to the gained recognition of required processes, its role as a guideline is weakened. Often, the final business process framework is not ready before the process modeling is completed. After the modeled processes are implemented or when presenting the organizational structure, the business process framework can then be ideally used as a communication means and as a navigator by a number of process models that were generated within the scope of the project.

The business process framework of DeTe Immobilien spans three different customer segments (Figure 4.11). First, the parent company of Deutsche Telekom is owner of the managed facilities and therefore requests that the interests of the owner are considered when making strategic decisions, or even sets own defaults. Sec- *3 different* ond, the same company group is also user of the facilities and *customer* therefore the key account for services of DeTe Immobilien. The *segments* third customer segment consists of institutional investors for which facility development services are rendered. The strategy of DeTe Immobilien tends to open the external market. The services are mainly rendered to external customer groups (as users and / or owners), which is reflected by the strong weighing of the marketing and procurement processes within the Business Process Framework.

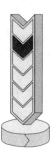

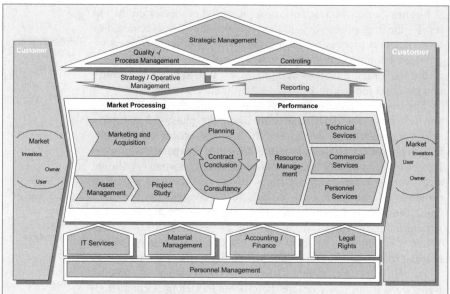

Fig. 4.11. Business process framework of DeTe Immobilien.

Core-, support- and guiding processes

The core processes of the company are managed by support processes. In the Figure 4.11, they are located below the core process level. Here, the "traditional" support processes are concerned, i.e., IT services, material management, accounting / finance, legal rights and human resource management.

Above the core processes, the steering processes can be found: quality / process management, strategic management and controlling. They communicate with the core processes via leadership and reporting.

Market exploration

When structuring the core processes, the elements of market exploration and fulfillment were applied and further developed. Market exploration mainly consists of two processes – marketing and acquisition. They focus on customers as users of facilities. In the process flow, they end with an interface to the planning and consultancy processes. The core processes of asset management and project study have another focus. Asset management reflects the owner's view on the asset portfolio. On the strategic level, the asset portfolio is managed by profitability. This may lead, for example, to the decision to convert an office building in a favorable shopping center. Thus, it is turned into a more profitable facility by changing its purpose of use and by developing it accordingly. Such a decision triggers the process project study which includes all the detailed planning for the re-development of this facility.

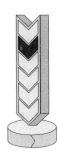

The linking element between market exploration and fulfillment in the business process framework is represented by the processes of planning, consultancy, and contract conclusion. The circular arrangement of these three single processes is the result of a long and intensive discussion and negotiation. This depicts another characteristic of a business process framework. It should be valid for the majority of the processes of a company. In particular, processes which fulfil equal functions and have primarily the same sub-processes but in different scopes, should be presented on the upper level in the same way. When undertaking planning services, customer consultancy, and contract conclusion, these three processes are arranged in a different way depending on the scope of services to be rendered.

Planning to be used as link

Standard products, e.g. a maintenance contract for heating equipment, only requires entering the type of equipment in the standard maintenance plan (planning), informing the customer about possible maintenance services (consultancy) and to complete and sign a maintenance contract (contract conclusion). If, however, a complex architectural or engineering service is concerned, e.g., the building of an office floor, then planning and consultancy are only performed at the time when the data of the offices is collected and when design alternatives are discussed. Then, the planning phase follows in which costs for reconstruction are calculated based on lump sum values. The related quotation is then presented to the customer, possibly negotiated and adjusted, and then forms the basis for the final contract document. After signing the contract, the initial planning needs to be revised and specified in more detail in order to be able to execute the contracted work.

Arrangement of contract conclusion

From this example, you can see that both processes are equal but, nevertheless, need a different arrangement of their partial processes when presented in serial sequence. In comparison, the contract conclusion process is the more advanced in that more planning and consultancy are needed, since the contract conclusion is followed by a comprehensive internal planning process. Since it is impossible here to show the many interim steps between start and end of the planning scope, the solution of a circular arrangement was selected.

Circular arrangement as solution

Fulfillment in the business process framework consists of processes with three different types of services offered by DeTe Immobilien: technical, commercial, and personnel services. They are preceded by the resource planning process which provides all three service processes with the resources required. Within the business process framework, the total value chain spans from customers to customers and represents, by its arrangement, the guiding principle for modeling the ensuing processes. Every employee is able to find

Performance

his field of activity and his position in the overall relationship of the processes.

A primary problem when creating the business process framework for DeTe Immobilien resulted from the different views of sales and service units to the internal control of these processes. While sales had, in the mean time, divided the products into 12 product families, the service units insisted on a separation of technical, commercial, and personnel services. All attempts failed to uniquely assign the product families, which spanned different areas, to the service areas. Controlling employees by categories, as defined by sales, was regarded as being impossible from the service people's point of view. This unclear orientation of the sales strategy was reflected in the development process of the business process framework. First, an attempt was made to map the 12 categories of sales, but because of the related complexity, the 3 service areas were depicted. In conclusion, the business process framework could only be finalized after the to-be modeling had been completed.

4.8
Checklist

What to observe!

Strategy and structure

- The strategy can only be derived from stringent ☑ structural changes. Therefore, be aware of the actual and future definition of corporate strategy prior to the modeling project.

- Most companies are structured by strategic business ☑ units. Evaluate if they are still valid and whether or not markets can be combined or have to be differentiated.

- Primarily, there are two market-oriented strategy types: ☑ Low-cost leadership and product differentiation. Avoid a "stuck in the middle" situation, because corporate structures need clear goals to which they must adapt.

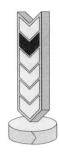

Core competencies

- Evaluate if an internal structure by core competencies ☑ could be of advantage for a climate for innovation and resource utilization.

- Identify your core competencies but be critical. Not ☑ everything that a company "can do pretty well" is a core competence.

- Abstain from old structures and, above all, from the ☑ actual organizational structure. In most cases, core competencies consist of a combination of capabilities and not of single capabilities themselves.

- Avoid polarizing the evaluation of organizational ☑ concepts. None of the concepts has a solution for all problems and every concept is based on a number of prerequisites which have to be revealed.

- The concept of efficiency criteria is ideally suited to ☑ define or to evaluate the direction and goals of the organization's structures. Think about the weighting of efficiency criteria to find out whether or not they are suitable for the related organizational purpose.

Business process framework

- The business process framework is more than just the ☑ public relations image of the company or the project. It will serve later as a means of communication and as a guiding principle for organizational changes. Therefore, keep the precise conformity of the processes and the continuous use in mind once the definitions are agreed upon.

- Refrain from using terms from the old organizational ☑ structure to define the new business process framework if you want to create something new. Otherwise, a references to old thinking patterns cannot be avoided – and nothing is stronger than sticking to traditional patterns of behaviour.

As-is Modeling and Process Analysis

Ansgar Schwegmann, Michael Laske

5.1
Objectives of as-is modeling

The identification of the core processes of a company is followed by the collection of detailed data related to these processes and an analysis of the currently performed processes. This information is the basis for identifying shortcomings and for recognizing potential improvements. For example, shortcomings, such as organizational interfaces along a process or insufficient IT usage, can be detected.

As-is modeling can involve considerable efforts. Therefore, the first question must be, whether or not, and in which scope, as-is modeling should be conducted.[113] The following reasons favor detailed as-is modeling.[114]

- Modeling the current situation is the basis for identifying shortcomings and potential improvements. Highlighting and explaining existing shortcomings facilitates "unfreezing" the organization.

Arguments for as-is modeling

- Sufficient knowledge of the current status is a prerequisite to developing a migration strategy to the new processes.[115]
- As-is modeling, in particular, gives an overview of the current situation for new and external participants in the reorganization project. This promotes the understanding of relevant relationships and existing problems of the company, and forms the basis for the design of adequate to-be models.
- As-is modeling can be used to train and introduce the project members to the tools and modeling techniques to be used. Any

---

[113] In the literature there is no consensus whether or not as-is models should be created. See Girth (1994), p. 147; Gaitanides, Scholz, Vrohlings, Raster (1994), pp. 257-258; Sharp, McDermott (2001), pp. 186-188.
[114] See Girth (1994), p. 148; Rosemann (2000), p. 22.
[115] See Gaitanides, Scholz, Vrohlings, Raster (1994), p. 258.

existing deficiencies of the techniques or tools will be detected in the preliminary stages of to-be modeling. This will enable the project members to concentrate on the improvement of structures and processes.

- An as-is model can be used as a checklist in the to-be modeling stage in order to prevent relevant issues from being overlooked.
- The as-is models can be reused as the main input for the to-be models, assuming that the current situation conforms, at least in partial areas, to the to-be models,. This will result in a reduction in the efforts for to-be modeling.

The following arguments oppose a detailed analysis of the actual situation of a company.

Arguments against as-is modeling

- The collection of data of the current status, i.e., as-is modeling, might limit the creativity of the participating employees. The danger of thinking in constraints is significant.
- There is a risk that old structures and processes will enter the succeeding to-be modeling phase without being carefully questioned.
- Creating detailed as-is models can be very time consuming and expensive. The required resources are determined by different factors. The expense increases, for example, when no consensus among the experts about the existing organizational structure and processes exists. Achieving consensus is all the more difficult when involving a large number of persons who have divergent or limited professional modeling know-how. In addition, the required expense is influenced by the scope of documentation of the actual situation and how current the existing documentation is.

In summary, it can be stated that as-is modeling – even if rudimentary – is always useful. "A key concept – we are seeking to understand the as-is, not document it in excruciating detail!"[116] Identified weaknesses and potential improvements serve as substantiated arguments for change to put to management, steering committees, and employees in order to motivate them for future changes. In addition, as-is modeling often enables at least parts of the detected shortcomings to be eliminated easily and some improvements to be realized quickly, prior to a comprehensive reorganization.

---

[116] Sharp, McDermott (2001), p. 186.

5.2
Procedure of as-is modeling

5.2.1
Preparation of as-is modeling

Preparing for the as-is modeling task requires a number of steps. First, it has to be determined which level of detail and which modeling techniques are to be used to create the models and which views are to be modeled. These factors are determined by the objectives, requirements and restrictions of the project.[117]

Select the level of detail

The level of detail to be selected for the as-is models is influenced not only by the modeling objectives, but even more so by the extent to which portions of the as-is model can be incorporated into the to-be models. The high expense of detailed as-is modeling can be justified only if it can be expected that a significant part of the as-is model can be transferred to the to-be model. Besides the level of detail, it has to be determined which views need to be considered for as-is modeling. The Architecture of Integrated Information Systems (ARIS) makes a distinction between five descriptive views: Organizational view, process (control) view, data view, functional view, and output view.[118] In order to improve the organizational structure and the business processes, the organizational view and the process view are of primary relevance. In addition to the selection of the relevant views or perspectives, view-specific and general modeling guidelines have to be defined. These form modeling conventions, which serve the method experts as guidelines for the model design.[119]

Select the relevant views

Create the modeling conventions

Another step in the preparation of as-is modeling is the identification of relevant sources of information. One potential information source are existing organizational manuals that describe the current organizational structure and process organization. Other valuable sources are the documentation of the application systems in use (e.g. in the form of data models, function models, requirements specifications, user manuals). The sources of information have to be reviewed and judged in terms of quality and relevance.

Identify the information sources

Of special importance for successful as-is modeling is the involvement of qualified experts who are familiar with the company and with the structures of the existing organization and business

---

[117] For possible project goals refer to Chapter 2.1.

[118] For the Architecture of Integrated Information Systems (ARIS) see Chapter 3.2.

[119] For modeling conventions see Chapter 3.4. See also Rosemann (1998b).

*Create a
project plan*

processes. Furthermore, existing documentation is unfortunately often not up to date and does not reflect the operational reality. The consultation of qualified employees is essential, since only these people can inform about the actual situation of the company. In approaching as-is modeling, the related activities have to be scheduled in terms of time and capacity. Finally, they have to be consolidated in a project plan.[120]

An important activity in the preparation of as-is modeling for DeTe Immobilien was the development of the modeling conventions and the draft of a comprehensive procedure for the successful execution of the project.[121] Additional activities included the recruitment of suitable project participants and the identification and evaluation of available documents.

*Select user
experts and
reference
departments*

DeTe Immobilien was founded by a merger with an independent daughter company of Deutsche Telekom AG and has a number of subsidiaries that are characterized by partially deviating structures and processes. These conditions made the selection of adequate reference departments and project members difficult, since "political" factors had to be considered. The managers of DeTe Immobilien had difficulty finding sufficient common consensus regarding which organizational structure and which business processes[122] are considered best practice and which subsidiaries and / or departments are to be used as a reference for the as-is modeling phase. Finally, for the purpose of as-is modeling, a subsidiary was determined as a reference subsidiary by the project management in agreement with corporate management. Important special cases were collected in other branch offices.

Another problem was that the selected experts were involved in the daily business and, therefore, were not fully available for as-is modeling. These restrictions had to be considered in the project schedule. A temporary relief of the experts' workload in favor of as-is modeling was normally not possible since these employees were primarily managers who could not be replaced by other employees. Because of the importance and necessity of the project for the company, these experts generally accepted the additional workload.

*Evaluate
existing
documents*

During the preparation, a number of different, potentially relevant documents were identified. These documents had been created in other projects by different people (consultants, supervisors, etc.) with different professional backgrounds and varying objec-

---

[120] For project management see Chapter 2.
[121] See Chapter 3.4.
[122] Best practice are the processes, procedures, organizational structures, technologies, etc. that are regarded as "optimized" at a certain time.

tives. In addition, the documents were partially obsolete, often based on different technologies and modeling techniques, and already featured to-be specifications. Consequently, the evaluation of those documents turned out to be difficult. Only a few documents could in fact be used for as-is modeling.

5.2.2
Identification and prioritizing of problem areas

The problem areas that should be considered for as-is modeling need to be identified based on the existing conditions (organizational structure and business processes, products and services, etc.). A recommendation is to roughly split the problem domain into partial areas, which can be better handled, prioritized and processed by the modeling team. This splitting should be applied according to the created business process framework and consider the identified core processes.[123] For splitting, various alternatives are possible. First, it is necessary to distinguish between a function-oriented and an object-oriented splitting of the problem domain. Then the extent to which this splitting of the problem domain should correspond with the structure of the organizational hierarchy must be analyzed.

Splitting types of problem domain

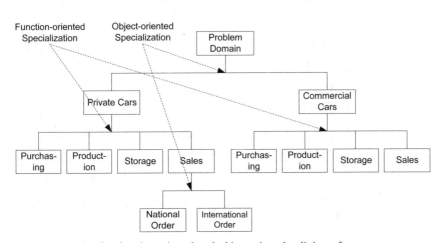

Fig. 5.1. Example of a function-oriented and object-oriented splitting of a problem domain

---

[123] For the creation of the business process framework and the identification of the core processes of a company, see Chapter 4.

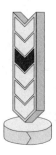

Function-oriented vs. object-oriented splitting

A function-oriented splitting includes the division into operational areas, such as purchasing, production, sales, accounting, etc. An object-oriented splitting includes objects that represent a group of similar products (e.g., commercial cars and private cars) or other distinguishable objects (e.g., documents, such as national and international invoices) which are subject to processing by operational core and support processes.[124] Normally, both splitting principles are applied in combination. Figure 5.1 shows an example of an object-oriented and function-oriented splitting of a problem domain.

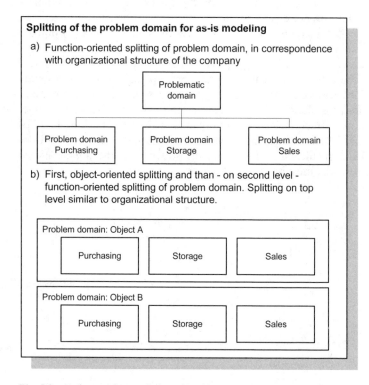

Fig. 5.2. Reference from splitting of problem domain to division of organizational structure

---

[124] An object-oriented organizational design can also be done based on input objects (raw material, interim products, services, etc.) or resources (tools, transportation units, etc.). See Kosiol (1976), pp. 50-51. For the object-oriented design of an organizational chart, see Chapter 7.

Object-oriented splitting of a problem area results typically in a stronger orientation towards the business processes as a process reflects by definition the flow of a relevant object through a system. Thus, the interdependencies along the entire process become visible. This is an important advantage in comparison with the traditional function-oriented splitting of a problem area. An object-oriented domain splitting with a simultaneous function-oriented organizational structure requires different user experts from different departments to be consulted in order to design complete process models. This increases the efforts for as-is modeling significantly since the coordination of the modeling teams and the agreement on the results can be complicated. In particular, insufficient coordination of the modeling teams bears the risk that redundancies result out of modeling similar structures and processes.

Division based on organizational structure

The division of the problem domain either orients itself on the actual organizational structure of the company or is totally independent from it. The advantage of a division that is based on the actual organizational structure is that the relevant user experts can easily be identified and organized in modeling teams. The structure of the problem area does not necessarily need to correspond with the organizational chart. A function-oriented organizational structure can very well exist along side an object-oriented division of the problem domain. Figure 5.2 gives an example of how the function-oriented organizational structure can be divided into different object-oriented problem domains.

Rough collection of technical terms and processes

Within the identified problem domains, existing processes and structures have to be identified prior to modeling in order to get an overview of the related domain. Technical terms and processes are, in this context, of special importance. Collecting technical terms is necessary in order to detect differences in the use of the terminology within the same company and to arrive at a standardized terminology for the as-is models. In addition, standardized technical terms serve as a source for creating a data model.[125]

It is important to perform a pre-collection of processes because this allows a differentiated view on the functions in a company and forms the basis for the identification of weaknesses. The potentially relevant processes must first be collected rudimentarily, and include details of different characteristic features. The following list contains a number of features that should be collected for every process:

Features to characterize processes

- name of process
- objectives of process
- whether the process is an existing or a planned process

---

[125] See Chapter 3.

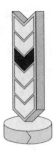

- whether the process is a core or supporting process
- the extent to which the process is documented, how it is documented and how current this documentation is
- documents and products within this process
- a possible process owner (name, organizational unit)
- participating organizational units (e.g. department, teams, positions, cost center, employee), the number of different organizational units and the related number of employees
- embedded application systems, databases and graphical user interfaces
- any contacts to external business partners (customers, suppliers, etc.)
- information about how frequently the process is executed
- the average processing time of the process and its variance
- the error frequency of the process (e.g., number of complaints or required reworks)
- the process costs
- the number of process participants, what need for reorganizations exists and how urgent such a reorganization is
- estimation by the method expert of how far the process can be incorporated in the to-be model, or whether this process is not suitable as a base for the to-be model at all

Criteria to prioritize the aspects of as-is modeling

It must be emphasized that the list of identified processes, which have been characterized, can only represent a rough guideline as to which aspects should be described within the as-is modeling phase. Whether or not the identified processes effectively become a part of the consolidated as-is model will become obvious when structures and processes are modeled in detail.

Not all areas of a company are equally relevant for as-is modeling. Areas which are classified as unimportant in view of the modeling objectives and which do not represent any core processes can be omitted in the as-is modeling. In addition, it must be examined whether special cases have to be mapped completely, or whether modeling only essential aspects can be regarded as sufficient ("80/20-Rule"). The selection of the areas to be modeled can be based on the identified processes and their characteristic features. In the following, some criteria are listed which enable a prioritization:

- Area represents current and / or future core processes[126]: core processes reflect the core competency of a company and the value-adding activities. Therefore, the as-is modeling of these

---

[126] For identification of core processes, see Chapter 4.

processes for establishing sufficient transparency gains significant importance. Even relatively unimportant processes can still be converted to future core processes because of a shift in the strategic orientation of the organization. Therefore, the future status of a business process needs to be considered when selecting process candidates for as-is modeling.

- Cost-consuming areas and / or processes: in addition to the value-adding core processes, the cost-intensive support processes with a high frequency of executions require special attention in order to detect potential cost savings.
- High need for reorganization: a need for reorganization exists if processes are executed ineffectively. This is the case, for example, when numerous process interfaces exist which cause long processing times and / or high process costs. A high demand for reorganization is also implied when inferior process quality leads to a high number of complaints and / or rework.

The list of prioritized areas and / or processes is the input for the collection phase and the documentation of the as-is models, as described below.

On the upper level, a function-oriented splitting of the problem domain was selected, which closely corresponds with the first level of the organizational structure of DeTe Immobilien. Within the functional areas, a partial object-oriented splitting of the problem domain was selected because the different objects showed some special features. *Splitting of the problem domain*

Parts of the identified processes have been renamed, aggregated, or split several times during as-is modeling. For this reason, a time-consuming, detailed collection of process features, using related project-form sheets, was only of little use. A rudimentary documentation was sufficient to get an overview of the areas to be modeled and to prioritize the processes according to the indicated criteria for as-is modeling. *Prioritizing the processes*

5.2.3
Collection and documentation of as-is models

In comprehensive modeling projects, the formerly identified and prioritized problem areas and processes have to be grouped prior to the creation of as-is models. A complex includes the as-is models to be designed by a modeling team. Every modeling complex *Modeling complex*

needs to be assigned a team leader, method experts, and competent user experts.[127]

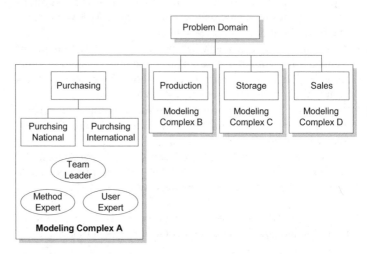

Fig. 5.3.   Splitting of problem domain into modeling complexes

Training and motivation of employees

Prior to starting the actual collection of data, it is necessary to motivate and train the project members. A sufficient explanation of goals, of possible consequences, and of the importance of the project to the company, is essential for the committed cooperation of the participating team leaders and user experts.[128] Otherwise, the danger that weaknesses are hidden exists because of a fear of change; facts are described incorrectly; and information is withheld.

Normally, the employees of the departments are not familiar with the modeling techniques and tools of conceptual modeling. Therefore, introduction and training for the project members is of vital importance for the acceptance of the new way of depicting organizational processes.[129]

The design of as-is models is best done in one or more workshops. Within these workshops, different aspects need to be considered:

---

[127] See Chapter 2.3.

[128] See Weidner, Freitag (1996), p. 273.

[129] For the selection of modeling techniques and tools, see Chapters 3.2 and 3.3.

- The typical goals of workshops are the modeling of relevant processes, organizational structures, rough functionality of existing IT infrastructure, core characteristics of related technical infrastructure (e.g. warehouse structure, means of transportation, production machines, etc). Normally, modeling is done iteratively. This means that the drafted processes are detailed and consolidated successively until a stable model is created which is deemed by all team members to be a suitable representation of the current status.

Organization and contents of workshops

- The technical terms being used in a company have to be identified, defined and consolidated. The semantic relationships between these terms have to be captured. Homonyms and synonyms need to be documented explicitly.
- Quantitative and time data of the processes must be collected as far as this is possible and at costs that can be justified in terms of the objectives of as-is modeling.
- Shortcomings and potential improvements need to be identified and documented in issue registers. Many of these issues can be easily specified by the involved users. As far as possible, reasons for the existing shortcomings and alternative solutions for improvements must be collected.
- Open points to be discussed in future workshops need to be presented. When executing workshops, questions may arise which require the involvement of other experts. Moreover, consensus of the workshop participants might not be achieved and a decision by the management might be required. Such topics have to be documented and processed by method experts and / or responsible supervisors.

Depending on the number of workshop participants, various workshop types can be differentiated:

- Single interviews by the method expert with one participant only or workshops with a limited number of experts from the modeling complex. In these meetings, critical subjects are discussed which exist in a modeling complex. Further, organizational problems are clarified (e.g. the selection of user experts).

Workshop types and their meaning

- Single interviews with user experts from the modeling complex. Here, selected aspects of the problem domain are discussed and modeled with highly qualified user experts. Single interviews allow for an efficient model design since long discussions with groups having differing interests are omitted. Another important advantage is that the user experts in single interviews are more willing to talk about weaknesses, which they would not do in group workshops because they would be afraid of negative reactions. Since the view of just one representative is mapped, the

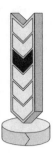

results from these single workshops need to be presented to the group workshops. Only in this way can a general acceptance of the outcome be achieved.

- Group workshops with the team leader, method expert, and all user experts of a modeling complex. The goal of this workshop is the collaborative determination of the as-is status and the approval of consolidated results.

Often, different opinions exist in regard to the contents and the graphical layout of the models. The discussion leader (a method expert), therefore, needs to conduct the discussion in such a way that the main goal of the workshop to produce an outcome remains the focus and long discussions of single aspects of the problem area are avoided. In addition, the discussion leader should limit the topic of to-be modeling to the identification of shortcomings and potential improvements. Shortcomings and / or potential improvements need to be documented in order to facilitate the analysis of the as-is models.

Post-processing of workshops

After the workshops, the models must be thoroughly edited by the method experts and presented to the workshop participants for approval. In the following workshop, existing deficiencies can be discussed once again.

The as-is modeling of a modeling complex is completed when a satisfactory degree of detailing and sufficient consensus regarding the achieved modeling results has been reached. Sufficient consensus is the aim as practical experience has shown that unanimous consensus among the workshop participants is virtually impossible to reach.

The creation of the as-is models within the project for DeTe Immobilien required approx. five months. In total, approx. 200 process models were collected and documented using the ARIS-Toolset. For the most part, generally valid processes were documented. The collection of variants and special cases were omitted.

Acceptance of modeling technology

The partial lack of acceptance of the selected EPC (Event-driven Process Chain) notation was problematic. The rejection was explained by the fact that another notation was already in use and also by the specification of EPC modeling, which led to irritations.[130] Another criticism was that seemingly trivial events that did not lead to any semantic enrichment of the models were included in many cases (example: the function "Create Order" is followed by the event "Order is created"). This, instead, caused a

---

[130] See Chapter 3 for the details of the EPC-notation.

considerable increase in the size of the model. In addition, EPCs alone are not sufficiently meaningful. Written documentation is in most cases still necessary if a third party has to understand the generated models.

In two modeling complexes, existing process documents were already available in the form of ABC flowcharts. These flowcharts were transferred to the EPC notation and agreed upon by the user experts and / or method experts who participated in the creation of the flowchart. A second round of agreements became necessary since the documentation in the ABC flowchart was not sufficiently meaningful and left room for different interpretations. However, the overall time for these activities has been rather short.

Transformation of existing models

In parallel with the as-is modeling of processes, the identified shortcomings and potential improvements were documented continuously. Additional interviews were held on site to identify the shortcomings of operational processes; such as the planning algorithms for staff involved in field services.

Identification of weaknesses

A detailed analysis of quantity and time frames for the modeled processes was omitted, since a rudimentary knowledge, in particular of processing times and execution frequency, seemed to be sufficient to gain an impression of the existing processes and to identify weaknesses. A detailed collection of meaningful parameters was not possible or would have caused unreasonably high costs in view of the large number of processes. In addition, the high degree of aggregation of the process models resulted in a great variance in parameters. For example: the time period for the planning process varied between one hour and six months.

Quantity and time frame

In summary, it can be stated that in this project the benefits from as-is modeling in relation to the costs have been rather low. A great number of the shortcomings could also have been detected without detailed as-is modeling. In addition, the created models were not suitable for reuse for to-be modeling. The reused models only amounted to approximately 20 percent.

Advantage of detailed as-is modeling

5.2.4
Consolidation of as-is models

To produce an integrated as-is model, all the models of the different modeling complexes need to be consolidated. In the following, different aspects will be discussed which are of importance for the consolidation of as-is models.

The modeling teams must already harmonize their models in the early stages of model design in order to facilitate later consolidation ("design for integration"). Therefore, the definition of

*Timely arran-
gement of
consolidation
activities*

*Consolidation
of object-
oriented
splitting*

modeling conventions and the standardization of technical terms (terminology) are indispensable.[131] In addition, a periodic agreement on the models in the modeling phase according to the selected splitting of the problem domain is necessary in order to facilitate a later combination of the models and to delimit the later comprehensive model adaptations.

If an object-oriented splitting of a problem domain is executed, an early analysis must be performed to determine whether or not equal structures and processes of different objects exist. These structures and processes have to be documented in a model directly. This will decrease the integration costs. An object-oriented splitting bears the risk of creating object-specific partial models, devised by different teams, and cannot be compared later because of the different structuring and naming of the model elements, or only at considerable cost. Object-specific partial models, therefore, have to be designed, if possible, with corresponding structures in order to support the comparability of the models. This facilitates the later analysis of the models for possible shortcomings and potential improvements. For example, a standardization of the processes of different objects can reduce the costs for development and adaptation of application systems.

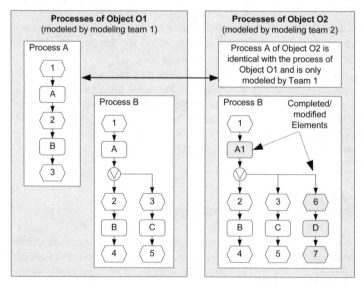

Fig. 5.4. Modeling of different objects with low redundancy and / or analogous structures

---

[131] See Chapter 3.

Figure 5.4 outlines the previously mentioned scenarios. Process A is identical for object O1 and for object O2. Therefore, it is modeled only once. Process B has a specification which differs for objects O1 and O2. In order to be able to compare the specific processes of these objects, the modeling teams have to assign equal names to the model elements, and, if possible, arrange them in the same way. This will lead to highly analogous model structures.

If a function-oriented division of the problem domain is performed, the main objective is to harmonize the interfaces between the modeled areas. An interface description documents which physical or digital objects in which specific state are transferred to the succeeding processes. The input- and output objects of the modeled problem areas such as purchasing, storage, or sales, and – in line with them – the specific tasks of the corresponding partial processes, have to be specified in detail and compared in order to achieve a consistent total model.

Consolidation with function-oriented division

In general, a uniform degree of detailing needs to be aimed at facilitating the comparability of partial models and to make a precise definition of interfaces easier. It can also be a good idea to use different levels of abstraction for the models. Partial areas with stable processes and structures enable a more detailed modeling on a lower level of abstraction when greater detail is required according to the aims of a particular model to be created. Problem areas that can only be presented in an abstract form are those that are subject either to a permanent change and do not include any stable structures or processes, or that are changed considerably during to-be modeling.

Consolidation of models with different levels of abstraction

After the conclusion of team modeling activities for the individual modeling complexes, the created partial models need to be integrated into an entire model according to the formerly described principles. Normally, this will cause inconsistencies that require the partial processes to be reworked and agreed upon again. The total model should be structured to correspond to the business process framework.[132] The business process framework serves as a top-level-model and is the starting point to navigate through the consolidated entire model.

Creation of an entire model based on the business process framework

Normally, a consolidated model will show great complexity because of the large number of process interdependencies to be considered. In order to more intuitivly understand the described procedures, it makes sense to explicitly model selected business transactions that are of special importance to the company, and to present them on the top level.[133] In contrast to the business process framework, control structures (alternative scenarios, loops) can

Modeling of selected business transactions

---

[132] For creation of a business process framework see Chapter 4.
[133] For presentation of business transactions, see Chapter 6.2.2.

also be modeled to a limited extent. In addition, color coding can serve, for example, to recognize the organizational units to which they belong.

Skeleton conditions of consolidation

For DeTe Immobilien the consolidation of the models of different modeling complexes turned out to be rather difficult since the harmonization of the models was performed relatively late. In addition, the models had been created on different levels of abstraction and differentiated by different objects. For example, in the modeling complex "planning and building" just one very abstract, generalized presentation of the processes was possible since the underlying projects differed considerably based on the individual scenario. Sales, however, enabled a relatively detailed presentation because of stable processes. Furthermore, there was no differentiation made by objects in the sales processes, while such a differentiation was possible in the areas of consultancy and service. For these reasons no complete consolidation was aimed at. This was not a problem for the purpose of as-is modeling since the model should primarily serve as an overview of the actual, current status.

Creation of an overview model

In order to create a compact modeled area, a high level overview model was created, which closely corresponded with the formerly designed business process framework. This high level model served as a starting point for navigation through the as-is models and was sufficient to get a good overview of the relevant problem areas. The presentation of business transactions for as-is modeling was omitted.

5.3
Analysis of as-is models

5.3.1
Criteria for the evaluation of as-is models

First, the objectives of a company must be determined in order to evaluate the analyzed as-is model. In general, the goals of a company can be divided into functional, commercial, and social goals.[134] Table 5.1 shows an example of a goal system that must be known to enable an organizational design of a company.

---

[134] For the formulation of goals, see for example Schulte-Zurhausen (1995), pp. 321-331.

Table 5.1. Goal system for organizational design (based on Schulte-Zurhausen (1995), p. 326)

Functional Goals	Financial Goals	Social Goals
Focus: Performance	Focus: Profitability	Focus: Employees / Groups
Examples: • Reduction of run-times • Increase customer satisfaction • Reduction of idle times • Reduction of error rate • Increase of product quality	Examples: • Decrease of personnel costs • Decrease of administration costs • Decrease of capital commitment • Increase of income	Examples: • Securing of workplaces • Ergonomic workplace design • Identification with the company • Personnel development

The following features have been determined in order to put these goals into operation and thus give guidelines for an identification of shortcomings and potential improvements.[135] These features are grouped into three categories. Of special importance are the categories "Information Technology-Support" and "Process Organization".

Operationalization of corporate goals

IT-support and technical infrastructure

Adequate IT support for the organizational structures and business processes is a critical success factor in modern enterprises.[136] Therefore, the as-is models must be analyzed for the following potential problems:

Evaluation of IT support and technical infrastructure

- Missing functionality in existing application systems.
- Missing or insufficient options to manage relevant data and / or administration of irrelevant data.
- Redundant storage of data in different application systems. This may require multiple entries that result in higher maintenance costs and the necessity to implement and maintain interfaces. In addition, redundant data bears the risk of creating inconsistent databases.
- Low performance of information- and communication systems.
- Poor operability and / or inconsistent user prompts. This leads to high training costs and erroneous entries.
- Use of different information- and communication systems for the same task in different company areas. This results in high administrative costs, incompatible interfaces and communication problems between the members of the areas involved.

---

[135] For potential shortcomings, which can be identified during as-is modeling, see Becker, Schütte (1996), p. 95; Schulte-Zurhausen (1995), p. 319; Eversheim (1995), p. 143; Krickl (1994), pp. 28-29.

[136] See Davenport (1993), p. 37.

- Insufficient electronic data exchange with business partners (e.g. orders, delivery notes, or invoices).
- Lack of applying "new" technologies such as workflow management systems[137], electronic document management, web services, etc.

Besides the IT-related inefficiencies, deficiencies may also exist in the technical infrastructure. At a logistical service provider, poor route planning may hinder an optimal disposition of trucks. In retailing companies, poor capacity load strategy in the store may lead to products with overdue expiration dates.

Process organization

Evaluation of process organization

A number of criteria exist for the analysis of the quality of the process organization which may point to potential shortcomings:

- Identification of obsolete processes. A considerable rationalization can be achieved by identifying processes that can be eliminated altogether. An elimination of these processes normally requires that another organizational solution can be found that makes these processes superfluous (e.g. order entry by customer using the Internet or outsourcing of certain tasks to external service providers).
- Identification of potential acceleration. The as-is analysis must identify unnecessary activities and potential for the simultaneous execution of activities. In addition, the processes should be examined to find out whether or not they can be accelerated more economically by the increased use of application systems.
- Localization and optimization of internal and inter-operational process interfaces. An important potential rationalization is the reduction of internal process interfaces. A process interface results when the processing of an object moves from one organizational entity (department, position, employee, etc.) to another organizational entity. The elimination of interfaces can reduce idle times and process initiation times. In addition to the internal interfaces, inter-operational interfaces to business partners can also be optimized. Example: Through the harmonization of transportation means, the processes of GR (goods receive) and GI (goods issue) become more efficient.
- Localization of workflows with same contents but different structures. The uniformity and standardization of processes facilitate the IT-support of these processes and, from the process manager's point of view, decrease the costs involved with the

---

[137] For the use of workflow management systems see Chapter 10.4.

introduction of new employees, as well as the complexity of the process.

- Elimination or redesign of forms. Too many forms or forms of poor design influence the efficiency of the processes.

Organizational structure / personnel

Deficiencies in the organizational structure and / or in the personnel area can be numerous and of very different kinds. Possible weaknesses are:

Evaluation of organizational structure / personnel

- Deficiencies through unclear, unsuitable and / or inconsistent assignment of decision- and processing responsibility.
- Unclear assignment of tasks from the customer's point of view.
- Too many hierarchical levels that participate in the decision and communication channels, prolonging decisions and preventing the employees from exercising their own responsibility.[138]
- Missing incentive systems to sufficiently motivate employees.
- Expecting too much or expecting too little of employees in their daily work.

5.3.2
Support of as-is anaysis by reference models

A reference model serves as a model for a generally valid documentation of best practice within a specific problem area.[139] In contrast to individual company models, being constructed according to the specific needs of that company, reference models are valid for a class of situations. Reference models often include a large number of partial model alternatives that reflect the different business events. They can be technical- or more business-related models. In the context of as-is modeling, business-related models, such as the retail reference model[140] or software-specific reference models,[141] are of importance.

The term "Reference Model"

A reference model is created inductively by the consolidation of know-how from existing models, application system documentation, conceptual specifications, expert consultancies, etc., or is deductively derived from theoretical frameworks. Within a company, the models from an important business area can be used as an internal reference for other business units.

Design of reference models

---

[138] See Eversheim (1995), p. 137.
[139] For the definition of reference model see Becker, Schütte (1996), pp. 25-26; Raue (1996), p. 26-27.
[140] See Becker, Schütte (1996).
[141] See Chapter 10.2.

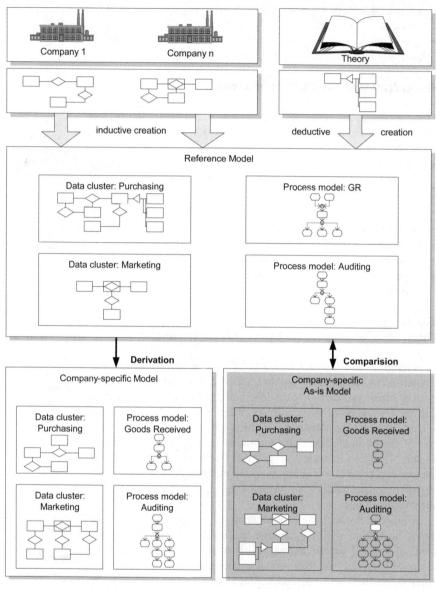

Fig. 5.5. Design and use of reference models (following Becker, Schütte (1996), p. 26)

Reference models are primarily used in two ways: They are used for designing or for comparative evaluations of company-specific models.[142] When used for designing, a company-specific model is derived from the reference model in that the know-how of best practice of this area is transferred to the company-specific model.[143] The use of reference models as a basis for modeling increases the modeling quality of a company-specific information model and accelerates the process of modeling as well.

Use of reference models

The advantage of using reference models in an as-is analysis is the easy access to knowledge that is documented in a similar notation. The model can also serve as a basis for a comparison of the documented as-is models. Reference models represent a neutral reference point to judge the quality of the as-is model and to identify weaknesses and / or potential improvements. There are, however, two problems. Firstly, the acquisition of adequate reference models is problematic and requires an investment. For example, up to now a reference model for the domain Facility Management does not exist yet. Secondly, a comparison of as-is models with reference models can be complicated if different modeling techniques have been used. The diversity of modeling alternatives for the same problem area often cause reference models and as-is models to be structured differently and to be modeled on a different level of abstraction.

Advantages of using reference models

5.3.3
Support of as-is analysis by benchmarking

The goal of benchmarking is the continuous comparison of parameters between company and company units in order to be able to judge the competitiveness of the company units.[144] The comparison includes parameters of products, services, processes, and procedures / technologies, to identify performance gaps to other external or internal company units that execute perceived best-practice in mastering certain processes or procedures.

Goal of benchmarking

Benchmarking requires first of all a specification of what has to be measured, e. g., products, processes or procedures.[145] Products can be marketable products as well as internal services (such as user support).

Select Measuring object

Every measuring object needs suitable criteria in order to allow a comparison of these objects. Important product criteria are costs, delivery time, quality or additional services. Suitable criteria for

Set measuring criteria

---

[142] See Schütte (1998), p. 309. For the advantage of reference models; see Becker, Schütte (1996), pp. 27-28.

[143] See Chapter 6.2.6.

[144] See Horváth, Gleich (1998), p. 326 and Chapter 10.6.

[145] For benchmarking of processes based on process cost calculation, see Horváth, Gleich (1998).

processes are run-time, failure rate, execution costs, etc. After these parameters have been set, they need to be calculated for the company. When calculating parameters, often a defect in the rating can be found, which means that the parameters cannot be calculated precisely or only at very high costs that would no longer be in a reasonable ratio to the benefits of benchmarking. In this case, the parameters must be estimated by competent experts.

Problem: Finding comparison partners

A key challenge of benchmarking is finding suitable partners for comparison because companies do not normally reveal their competition-critical data. When gaining direct access to company-critical parameters turns out to be impossible, the comparison can also be delegated to neutral third parties, in particular consultants with knowledge of different companies. In addition, it is important to consider whether or not a company-internal benchmarking in different company units is reasonable (e.g. subsidiaries).

Problem: Comparability of parameters

A further problem is the insufficient comparability of parameters. Parameters are calculated under different conditions (company size, customer structure, products, etc.), which must be observed when interpreting the parameters.

On the basis of the calculated parameters, the deviations between the values need to be evaluated. If significant differences are stated, the reasons for these differences must be identified and, if appropriate, suitable actions taken.

Compare parameters and derive parameters

Benchmarking can supplement the documentation of structures and processes within the scope of as-is modeling to identify weaknesses. Benchmarking, however, is only fully effective when the comparison of different objects, criteria, and partners is performed on a regular basis. Benchmarking should, therefore, be used in addition to the as-is analysis and / or to-be modeling. Within the scope of a reorganization project, it can also be used as an instrument for continuous process management.

5.3.4
Identification and documentation of weaknesses and potential improvements

Importance of as-is analysis

The primary goal of as-is modeling is the presentation of the existing structures and processes in a company. In addition, only known and obvious weaknesses and potential improvements are documented. The goal of the as-is analysis is to create a complete list of weaknesses and potential improvements based on the collected models. For this purpose, the as-is models are evaluated based on the previously discussed criteria. The process of localization of weaknesses and potential improvements is of a creative nature and requires, above all, experience and analytical capabilities.

In addition, reference models and the benchmarking results can be used to identify weaknesses and potential improvements. The assessment of identified weaknesses and potential improvements and their meaning for the company will be limited to a primarily verbal argumentation.[146] Normally, a precise monetary evaluation of weaknesses and / or potential improvements is not possible because the exact data that is needed can only be gained by expending an unreasonable amount of time and cost.

Localization of weaknesses / potential improvements

A systematic collection of identified weaknesses and potential improvements involves recording the following suitable information

Systematic collection of weaknesses and potential improvements

- ID Number for unique identification
- Short description
- Description of weakness, including possible reasons and / or description of potential improvements
- List of organizational units involved
- Classification (e.g. a weakness can relate to the organizational structure, to the process, or to the IT infrastructure)
- Meaning for the company and urgency
- Outline of alternative solutions
- Description of possible immediate actions to (partially) eliminate the weakness and / or to realize potential improvements.

The list of weaknesses and potential improvements is worked out by the method experts and user experts in different modeling complexes and subsequently consolidated by the method experts.

In the case of DeTe Immobilien, the majority of the more than 80 single issues had already been localized within the scope of as-is modeling so that in the analysis phase only a systematization and specification was required. A classification of weaknesses and potential improvements was done with respect to the relevant organizational unit and to the type of weakness and potential improvement. The following types were differentiated:

Importance of as-is analysis

- interface problems, pointing to poor coordination of different organizational units,
- insufficient IT support, and
- other organizational deficits (e.g., missing petty cash for service technicians).

Types of weaknesses

A comparison of the created as-is models with a reference model could not be conducted because at that time no reference models

---

[146] For organizational solutions, see Schulte-Zurhausen (1995), p. 447; Weidner, Freitag (1996), pp. 309-319.

Importance of reference models and benchmarking

existed for the problem area "Facility Management". However, for operational standard processes (such as accounting, HR-management) the SAP R/3 reference model was used to efficiently create processes such as invoicing and reminders. Because of missing comparable parameters and because of the high cost of comparable parameter collection, benchmarking was omitted.

5.3.5
Execution of immediate actions to eliminate weaknesses

Not all detected weaknesses require comprehensive reorganization efforts. Often, improvements can be achieved with only minor organizational actions. The following criteria may serve as pointers for situations where immediate actions could be taken and at what time in order to be successful:

- No change or just small structural changes to processes
- No adaptation or just minimal adaptation of existing IT infrastructure
- No new technical infrastructure
- No approval of measures by staff committee required
- Consensus among the participating managers and employees

Execution of immediate actions

A high number of identified weaknesses for one process is a trigger for immediate action. The manager should name one employee as responsible for planning, executing, and controling these immediate actions.

The following example demonstrates a reasonable immediate action that resulted in considerable cost savings for DeTe Immobilien: When buildings are constructed and the warranty period expires, the building must be inspected prior to the expiration date. In the past, this inspection had been done prior to warranty expiration, but in some cases this was not done in time to initiate related legal steps. Subsequently, an immediate action was made to set the inspection for an earlier date and so adapted to the needs. Because of this simple organizational action, such failures – as well as losses from expired warranty claims – were avoided. An example of an issue register list is shown in Table 5.2.

Table 5.2. Example for an Issue Register

No.	Problem Description	Idea of Improvement
53	Unclear competencies for the tasks "Safeguard Owner Rights" and "Carrier of Public Concerns".	Define unique competence for process "Safeguard Owner Rights and Obligations".
64	Customer DTAG should inform service about the importance of certain assets. DTAG- and DTI-interfaces are not sufficiently defined	DTAG- and DTI-processes have to be consolidated. Contact from Org / IT to TEMPO and process owner of TEMPO-process 14b. Modeling of total process is provided.
89	Portfolio planning is hindered by non-transparency of DTAG-General Ledger	Consolidation process with DTAG to allocate General Ledger Accounting (book value, purchase value, annual depreciation)
91	Portfolio planning process is difficult because of poor and limited access to SAP R/2 RA.	Consolidation process with DTAG to improve access to SAP R/2 RA.
47	Approval of budget for building measures comes too late.	In-time approval of budgets

5.4
Checklist

Preparation of as-is modeling

- Define the goal for as-is modeling. What is the purpose ☑ *What to observe!*
  of as-is modeling?

- Select the relevant views in your modeling architecture ☑
  and determine the degree of detailing of the as-is models.

- Identify potential information sources and evaluate ☑
  them in terms of relevance and how current they are.

Identification and prioritizing of problem areas to be collected

- Describe the problem area / domain or processes based on adequate characteristic features. ☑

- Keep a glossary that contains the definitions of relevant technical terms. ☑

- Prioritize the problem areas / domains based on the characteristic features, and determine the areas / domains to be considered in as-is modeling. ☑

Collection and documentation of as-is models

- Group the problem areas / domains to modeling complexes to be modeled by a team. ☑

- Train the project participants in the modeling techniques and tools. ☑

- Model the relevant processes and structures in single interviews and / or in group workshops. Pay attention to depicting equal subjects with analogous structures. Also, make sure that the terms in the glossary are used consistently and up-dated periodically. ☑

- Follow the modeling conventions when modeling. ☑

- Document the obvious weaknesses and potential improvements, and avoid long discussions about the planned state. ☑

Consolidation of as-is models

- Work out analogous structures for different partial models, and standardize the terminology used in the models. ☑

- Integrate the models created by different teams into a total model at an early stage. ☑

- Structure the total model according to the business process framework and model typical business transactions, if possible. ☑

Analysis of as-is models

- Set criteria for the evaluation of the as-is models. ☑

- Check whether or not reference models for the related ☑ areas / domains exist, and use them as a basis for the evaluation.

- Analyze to what extent process benchmarking is re- ☑ quired and possible. Initiate corresponding actions.

- Identify the weaknesses of the actual status based on ☑ the evaluation criteria, and document these sufficiently.

- Check to what extent identified weaknesses or poten- ☑ tial improvements exist that can be realized with a minimum short-term effort. Initiate and supervise the elimination of weaknesses or the implementation of the potential improvements.

To-be Modeling and Process Optimization

Mario Speck, Norbert Schnetgöke

6.1
Objectives of to-be modeling

To-be modeling is based on the as-is models and on the issues that have been identified during the process analysis. The results of to-be modeling and of evaluated to-be processes are linked with company-internal expectations from management as well as from the employees. Among these internal expectations are:[147]

- Increase in profits
- Cost savings
- Streamlining of processes
- Reduction of planning times
- Shortening of processing times
- Information that is more up-to-date
- Better communication between company units via defined interfaces
- Minimization of idle times

Internal expectations

The external customer- and / or market-oriented requirements of to-be processes include:

- Higher process quality and resulting product quality
- Closer proximity to customers and better customer commitment
- Faster communication with market partners
- Higher process transparency for the customers
- Larger market shares; for example, through a faster response to market developments

External expectations

These expectations from to-be modeling need to be discussed with all employees involved in the modeling project, so as to avoid pro-

---

[147] See Stahlknecht, Hasenkamp (1997), p. 259.

ject members adopting false or negative expectations. Such negative expectations are: excessive cost savings demanded by management; or employee fears of an alteration to familiar processes or a reduced flexibility in the workplace. First, the model users (user departments and management) determine their targets (e.g., increase in customer commitment by 20 percent within the next year). Second, the purpose of to-be modeling can be derived (e.g. increase in customer commitment by certification pursuant to DIN ISO 9000ff.).

Motivation of participants

Motivation of all employees participating in to-be modeling, together with the consequent support by the management, need to be regarded as critical factors in to-be modeling. A transparent presentation of the relationship between the different expectations from business process modeling, and the relationship between the different model users and the targets of to-be modeling, is an important factor to be rendered by the project members during to-be modeling. The identified and determined targets need to be agreed upon with the model users, and their necessity and / or value must be documented. This will contribute considerably to later acceptance, and, finally, to the quality of the generated to-be models.

Objectives of DeTe Immobilien for to-be models

In the beginning of the reorganization project, DeTe Immobilien aimed at quantitative as well as qualitative goals. This included, above all, decreasing the process costs, shortening the processing times as well as improving the product quality and increasing the flexibility and transparency of processes. In total, the to-be processes needed to improve the market attractiveness of the product line.

The DeTe Immobilien project team followed the guideline that these objectives should be converted using new and changed business processes, a process-oriented organizational structure, and adapted information systems. The use of the Enterprise System SAP R/3 in large areas of the company was already decided on in the early stages of to-be modeling. The process models, therefore, needed to be suitable to support customizing activities. During the reorganization project, SAP implementation projects that had already started had to be supported, and future possibilities of introducing Activity-based Costing had to be taken into account.

6.2
Procedure of to-be modeling

6.2.1
Preparation of to-be modeling

Analogous to as-is modeling, to-be modeling requires an appropriate degree of detailing to be defined for the models. The related valid criteria, however, distinguish themselves from those of as-is modeling. To-be models must reach a degree of detailing that is of value and relevance for the employees involved in the process in order to be able to judge their impact on the organizational structure[148], on the activities[149] to be executed, and on the communication between the processes. As a consequence, to-be modeling may require a more extensive detailing than necessary for as-is modeling.[150]

Define degree of detailing

The descriptive views of the ARIS-architecture need to be derived from the sub-targets. The degree of detailing of individual views also needs to be defined.[151] The implementation of a workflow management system, or the development of IT-systems, normally requires a higher degree of detailing than a model that is used to design a new organizational structure. A suitable definition of the degree of detailing is evident, in particular, with respect to the high costs of model creation and maintenance.

ARIS-views

Table 6.1 gives an example of how an appropriate level of detail can be achieved. For every descriptive view, process-, organization-, function-, and data view, an example is given as to which degree these should be used for modeling. Here, it has to be considered that the defined degree of detailing may refer only to single partial models and / or domains of the total process model, depending on the related contents. The objective of implementing an Enterprise System for order processing, for example, will lead to very detailed models in this company unit. This degree of detailing will not necessarily be the same degree of detailing in other business processes, such as marketing. That is why the degree of detailing must be oriented not only toward the purposes of to-be modeling, but also to the domain in question.

---

[148] For precise information on how to derive a process-oriented organizational structure from the created to-be models, see Chapter 7.
[149] See Chapter 4.
[150] See Scholz (1993), p. 83. and Chapter 5.1.
[151] Chapter 3.1 already handles the required degree of detailing and the selection of relevant views. See also Scheer (1998a), p. 4; Scheer (1998c), p. 21.

Relevant views and detailing depth for the purpose of...

In the following, we will discuss the relevance of the four ARIS views as well as the adequate detailing depth for selected purposes.[152]

Organizational documentation

... organizational documentation

Within the organizational documentation, process models are of primary interest since the transparency of the business processes represents an important goal of this purpose. The organizational documentation reflects the interaction of business processes. The documentation has to provide the information regarding who executes which activities. Since the organizational documentation also contains management relationships, the organizational structure with the relationships between the individual organizational units has to be considered as well and maintained in the organizational view of the ARIS-models.

Process-oriented reorganization

... process-oriented reorganization

More detailed models are required for reorganizing existing business processes, since analyses and evaluations have to take place on a detailed level. In contrast to a pure organizational documentation, qualitative and quantitative parameters are suitable for evaluating and reorganizing business processes. Therefore, models should be maintained that provide information regarding processing times, costs, or required changes.

Continuous process management

Continuous process management is an ongoing task and a continuation of the process-oriented reorganization, featuring the same requirements as the models created in as-is process modeling. Continuous management further requires that various cost and time attributes are assigned to the processes and functions.

Certification pursuant to DIN ISO 9000ff.

... certification

To a large extent, successful certification is often based on good organizational documentation of the company. Therefore, the models of the organizational documentation are a suitable basis for certification.

---

[152] See also Chapter 3.1.2.

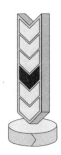

Benchmarking

The comparison of different company-wide or company-internal indicators requires a detailed description of performances and costs of functions. In addition to the qualitative and quantitative parameters, transparent process structures are the prerequisites for the use of processes of equal contents when comparing process performances and costs.

... bench-marking

Knowledge management

The use, generation, and distribution of know-how as a company resource can be traced transparently using detailed process models. The process models themselves serve as the "organizational memory" of business processes. The mapping of the knowledge status and the know-how needed by the organizational units and / or members contribute, for example, to a better design of end user training courses.[153]

... knowledge management

Selection of Enterprise Systems

The comprehensive integration of data and functions in modern Enterprise Systems (a.k.a. ERP, Enterprise Resource Planning) requires that a company compares the degree of coverage of their own data- and functional / process models against the degree of coverage of models that are supported by the Enterprise System. In addition, Enterprise Systems often offer the possibility of visualizing the organizational- and process models that are integrated into the software. The degree of coverage that is identified here can be taken as one measure for the suitability of the Enterprise System, and can give a first indication of the required efforts for the implementation of the software. The to-be models to be created should correspond to the degree of detailing of the Enterprise Systems models. The Enterprise Systems models may also serve as a basis for the enterprise-individual business models. This is especially suitable in standardized areas, such as accounting.

... selection of Enterprise Systems

Model-based Customizing

Some Enterprise Systems support a process of software individualization through a modification of the software's reference models. The availability of these models is advantageous because this can reduce the implementation costs. The achievable degree of detailing, however, widely depends on the capability of the selected

... model-based customizing

---

[153] See Chan, Rosemann (2002) and Rosemann (2001) for further details.

Enterprise System and, therefore, has to be determined prior to modeling by evaluating this solution.

Software development

... software development

The utilization of the models in the requirement engineering phase of software development requires a formal presentation with a high degree of detailing. The focal points of the traded software development methods, such as Structured Analysis (SA), are function- and data modeling. Also relevant for integration in the organizational structure and business processes are organization- and process models. Therefore, a model-based software development makes great demands in terms of formality and detailing of business process models.

Workflow management

... workflow management

Workflow management systems facilitate a (semi-) automated execution of business processes based on workflow models. In particular, the models of organizational-, data-, and process views must be modeled in detail. Process control can only be executed based on single attributes and the available information for single organization members has to be mapped.

Simulation of processes

... simulation of processes

Process simulation requires a detailed consideration of functions, processes and the participating organizational units. The functional view delivers time- and cost information that is used within the simulation, while the organizational structure delivers information about personnel capacities (working hours, etc.). Therefore, this data has to be collected in an appropriate level of detail. From the data view, only the data that is relevant for decision making (e. g. possible occurrence, execution frequency) has to be modeled.

Table 6.1. Weighing of descriptive views[154]

Purpose \ View	Functional View	Organizational View	Data View	Process View
Organizational Documentation	Low	Medium	Low	High
Process-oriented Reorganization	Medium	High	Medium	High
Continuous Business Process Management	Medium	High	Medium	Medium
Certification pursuant to DIN ISO 9000ff.	Low	Medium	Low	Medium
Benchmarking	High	Low	Low	High
Knowledge Managment	Medium	Medium	*	Medium
Selection of Enterprise Software	Medium	Medium	Medium	Medium
Model-based Customizing	Medium	Medium	Medium	Medium
Software Development	High	Medium	High	Medium
Workflow Management	Low	High	High	High
Simulation of Processes	Medium	Low	Low	High
Activity-based Costing	Medium	Low		Medium

Degree of Detailing   ⬤ High   ⬤ Medium   • Low

* When process models are used in knowledge management, no data or technical terms are modeled, except for the know-how which is required to execute the functions.

---

[154] For the individual purposes of process modeling, see also Chapter 3.1.2.

... Activity-based Costing

Activity-based Costing

The implementation of Activity-based Costing requires detailed functional models with cost, time, and quantity data. Organizational charts primarily provide cost records for employees and cost centers. Process models are relevant since they show which functions are executed in the process. Corporate data that is exchanged between the functions, however, need not to be considered in Activity-based Costing. Therefore, the necessity to create data- and / or technical term models in this special case is omitted.

Combined objectives

If to-be modeling has to cover multiple purposes, then the maximum degree of detailing of all objectives is required for every view to be modeled. However, a general degree of detailing cannot be determined for all business processes because it differs in domains depending on the purpose. Normally, the implementation of a workflow management system makes sense only for a part of all business processes. Therefore, only the relevant processes require a high degree of detailing.

The required capacities and times for the process phase of to-be modeling can be planned, building on the results of strategic planning where the core processes have already been identified,[155]. When planning the time, it must be observed, in particular, that the time schedule for the support processes is modeled in conformity with the core processes. The reason is that the requirements of support processes and / or related performances arise from the core processes.

It must be emphasized that the knowledge gained from as-is modeling may lead to an extension or modification of the modeling conventions for to-be modeling. Therefore, the original conventions have to be thoroughly verified and re-evaluated for to-be modeling.[156] Last, but not least, the understanding of the used, semi-formal model types for to-be models is widely dependent on these conventions since to-be models, in particular, are accessed by a vastly larger user group than as-is models.

Objectives at DeTe Immobilien

The primary goals of DeTe Immobilien for to-be modeling were as follows:

- Creation of a process-oriented organizational structure based on process models.
- Support of certification pursuant to DIN ISO 9000ff.
- Organizational documentation to support employees in the implementation and conversion of new processes.

---

[155] See Chapter 4.
[156] For modeling conventions see Chapter 3.4.

The goal with the highest priority here has been the creation of the process-oriented organizational structure. Because of the tight timeframe, the first consideration was to develop the organizational structure simultaneously with the to-be processes. However, the extreme complexity of single tasks and the great dependency of the organizational structure on the to-be process models gave reason to select a gradual approach.

The degree of detailing for all views of the ARIS-architecture was derived from the targets. The modeling was focused on the process view. The degree of detailing was selected to avoid a further refinement of automated functions. Manual activities of employees, however, were detailed according to the changed qualification requirements that varied from one function to the next.

The data / technical term models consisted of technical terms already determined and documented in the as-is models. The changes in the processes had little influence on the technical term definitions and structures. The technical term models, however, were recreated intentionally for some core terms such as "order". A further detailing of technical term models using data models was not intended in the first step, since the required costs seemed to be justified for software development purposes only.

The organizational view was not modeled because the new organizational structure was developed later, based on the to-be models. Also, for economic reasons, most of the functional views were not modeled.

6.2.2
To-be process identification and rough draft

A company consists of a number of constitutional processes with a multitude of cross connections. The core challenge is to identify the processes that represent the main activities of the company and to separate them from other processes.

In addition to the core processes of a company,[157] which are executed to provide related market performance, so-called support processes exist. A characteristic feature of support processes is that they have no direct relationship to the company-external market. This means that only company-internal demands for performances exist without any impact on the external market. However, disturbances in support processes may lead also, after a certain time, to disturbances of the core processes and are, therefore, indirectly relevant for delivering high value services to the customers. Examples of support processes are: financial management, human resource management, or IT services.

Core and support processes

---

[157] See Chapter 4.6.

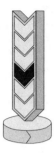

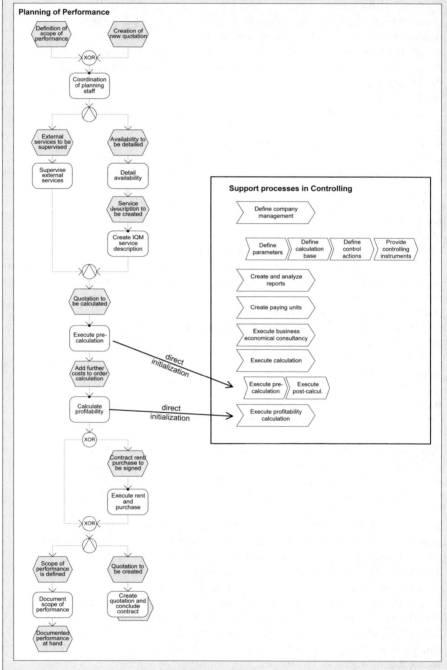

Fig. 6.1. Direct performance relationship between a core- and support processes

The interrelationships between the core processes and the support processes can be divided into direct and indirect relationships. A direct relationship identifies the performance exchange between support processes and core processes on the process structure level (type level) and can therefore be modeled. An example is the legal support (a support process) in the case of a missing payment and the related processes of generating and sending reminder notices (within the core process). The performance can be precisely localized in the process model. Figure 6.1 gives an example based on DeTe Immobilien.

Direct relationship

An indirect relationship between core processes and support processes cannot, however, be mapped in a process model, since either no direct performance exchange exists, or an identification of this relationship on the process structure level – i.e., abstracted from individual business transactions – is impossible. For example, identifying all interfaces between core processes and a legal support process is not practical because a legally relevant situation between the market partners may occur with every interface or with every performance exchange. An explicit modeling of every individual subject would overload the process model with non-relevant information and make it difficult to read. Consequently, it would become unusable. As a rule of thumb, approximately 80 percent of the process instances – i.e., individual business transactions – in the process models should be collected and modeled. Moreover, the high degree of detailing of most application cases contradicts the guideline of economic efficiency.[158]

Indirect relationship

80/20 rule

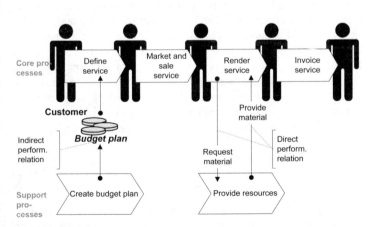

Fig. 6.2. Performance relation between core process and support processes

---

[158] See Becker, Schütte (1996), p. 65. and the introduction to the Guidelines of Modeling in Chapter 3.4.1.

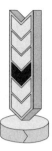

Even if no direct relationship can be identified, there is still the possibility of identifying an indirect performance exchange. An example is finance and cash flow management. Individual instances of the core processes require the services of the support processes, but do not necessarily initiate the related support processes because a detaching mechanism, such as budget control, is placed in between. Figure 6.2 gives an example of the performance relationship between core- and support processes.

In order to identify the relevant business processes for to-be modeling, the core processes in a company must be identified and sufficiently separated from each other as well as from the support processes. In principle, two methods can be used: top down and bottom up.

Top-down method

Based on the corporate strategy, the top-down method generates core processes from the strategic business fields.[159] These processes, on the highest level of abstraction, are successively refined in the course of modeling. The advantage of this method is the development of business processes, which are very close to the strategic viewpoints. The hierarchical refinement of process structures may cause a lower overall process performance for local processes. This is particularly the case when the resulting interdependencies between the partial processes are not considered or not recognized.[160] Often, the reason is the ignorance of conflicting resources within concurrent processes.[161]

Bottom-up method

The bottom-up method is based on the entirety of all planned activities. For every identified activity, process models are generated, from which the process structures on higher levels are derived through grouping.[162] The business processes are later divided into core- and support processes. The advantage of this method is the ability to create to-be models within a short time, especially because of the close connection to the operative core processes. Restrictions, such as existing IT systems, which need to be incorporated, can be added as desired to the created model in a timely manner. However, the advantage of a close connection to the operative activities of the company can also be considered a disad-

---

[159] See Sommerlatte, Wedekind (1989), p. 57 and the explanations in Chapter 4.3 on how to create a company-specific business process framework based on the corporate strategy.

[160] A detailed explanation of the top-down method is given by Remme (1997), who deals, in particular, with the modeling of process particles (reference process modules). Individual business processes are created by forming variants from process particles. This topic is further discussed by Scheer (1998c), p. 7, and Gaitanides (1983), p. 23.

[161] See simulation of business processes in Chapter 6.2.4.

[162] See Gaitanides (1983), p. 64.

vantage, since the individual activities are embedded in the business process models at a very early stage and can, therefore, only be evaluated later in connection with interacting business processes, regarding their relevance, necessity, and advantages.[163] In addition, the complexity to be mastered is increased by the high number of single activities that must be processed without reference to each other. Another potential problem is the necessary, complete identification of all corporate functions with the risk of overlooking relevant business processes already in this phase.

The rough draft of the to-be model should show the following results:

- Division of all identified processes into core- and / or support processes and their performance relationships. Here, the existing business process framework can provide valuable guidance.

Results of rough draft

- Rough process structures on the higher levels have to be created for modeling, i.e., a sequence of functions, or value chains only.
- A first group of core- and / or support processes has to be created with modeling complexes that must be as homogeneous as possible. The modeling complexes have to consider the total scope of performance, and they must be created in such a way that the number of process links between the complexes is small, but large within a complex. The process link is defined here by the number of interfaces and by the relationship between the processes and contents of this relationship. The function- and / or object-oriented procedures from as-is modeling can be used as principles for grouping and for hierarchy generation.[164]

From the defaults of business domain planning and strategic corporate planning, the following core processes for DeTe Immobilien were defined:[165]

- Sales

Core processes of DeTe Immobilien

- Consultancy
- Asset management
- Planning and building
- Service
- Facility development

The following processes were identified as support processes:

Support Processes of DeTe Immobilien

- Information processing
- Legal support

---

[163] See Küting, Lorson (1996).
[164] See Chapter 5.2.2.
[165] See Chapter 4.6.3.

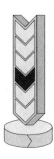

- Invoicing
- Controlling
- Personnel development and service.

The individual core- and support processes have been explained in a detailed description. An example of such a description is given in Table 6.2.

Table 6.2. Components of the core process sales

Gaining customers (sales promotion)	• Execute periodical calls • Initiate customer contacts • Plan fair visits • Select target groups • Visit customers • Calculate customer needs • Recognize potential customers • ...
Advising of customers	• Take stock of inventory • Create rough concept • Present solutions • Initiate consultancy about application, asset-management as well as planning and building • ...
Create quotation	• Collect quotation contents • Present quotation • Revise or modify quote • ...
Negotiate contract	• Build up trust • Execute / initiate ratification • ...

Based on strategic defaults, the cooperation between the different modeling teams had to be agreed upon. For this purpose, a number of workshops were held with all process godfathers and with the first expert users. The intent was to identify the business processes of DeTe Immobilien that followed an object-oriented division in contrast to a division into core- and support processes:

- Planning and execution of Im^2 [166]
- Maintenance of object value
- Development of products
- Continuous optimization of Im^2
- Provision of investment objects
- Portfolio management

---

[166] Im^2 is a product name of DeTe Immobilien and stands for "Intelligent square meter". This includes the rental of surfaces and value-added services such as rental or leasing of telecommunication equipment, disposal lines, etc.

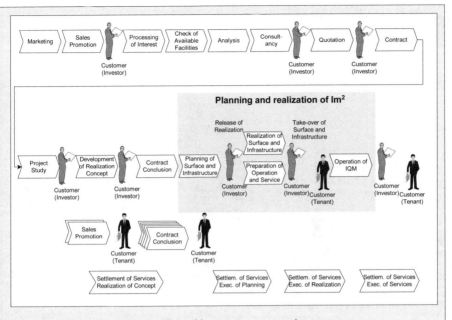

Fig. 6.3. Provision of investment object without core process assignment

In order to get a process-oriented division of the processes and to identify and / or define the interfaces between the modeling teams, every business transaction mentioned above was provided with a value chain similar to the specification of the core process. Figure 6.3 gives an example for business process "Provision of Investment Object ..." which used business the transaction "Planning and Execution of Im^2". Analogous to the description of single functions in core processes, the workshops, designed to identify the partial functions of every business transaction, determined their functional contents as well as their input and output operations (refer to example in Figure 6.5).

The clarification of the business process structure was followed by the identification of process responsibles for partial areas of the business transactions (Figure 6.4). The so created delimitation, description, and assignment of partial functions of a business transaction served as a basis for further education and eventual reformation of the modeling teams, as well as to coordinate and define the related complex to be modeled. This approach secured not only the availability of professional know-how for the modeling teams who had to create adequate to-be models, but also led to interactions between the core processes and the business transactions with the lowest possible inefficiencies

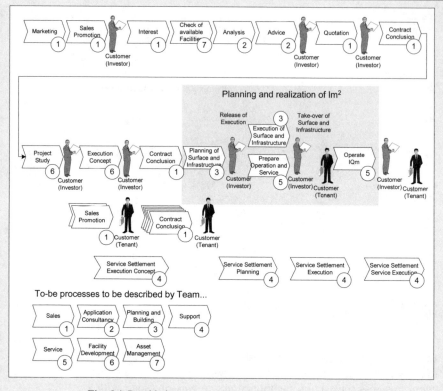

Fig. 6.4. Provide investment object including core process assignment

Business process: Provide and operate Im^2
Function: Create quotation

Input:
- Finished (fine) concept

Partial functions:
- Incorporate fine concept in quotation
- Determine the prices
    - Determine the covering margin
    - Calculate price margin
- Write quotation

Output:
- Quotation

Fig. 6.5. Example of a function description

Since the interfaces had already been clarified in the early stages, the later reconciliation actions could be clearly shortened and redundant tasks could be avoided.

This procedure contributed to a wide consensus of the most important business processes of DeTe Immobilien. The responsible persons who were already familiar with process modeling, largely assumed the role of communicating this process and the targets of process modeling for the total company. This broad basis was a necessary prerequisite for the generation of the single to-be models to follow.

6.2.3
Design and documentation of to-be models

The to-be processes are intended to operationalize the strategic targets of the company. Therefore, the extent to which all partial processes and single activities in a company contribute to these targets needs to be examined. Partial processes and / or activities which are not identified as value-adding processes and activities, and which do not serve at least one non-monetary corporate goal, must be eliminated from the business processes. Here, however, it is important to take company-external influences into account, because legal prescriptions, such as guaranteed data protection, have to be observed. As-is modeling is followed by the as-is analysis. The as-is analysis serves as an important guideline for to-be modeling. The identified issues must be investigated individually and be considered in the to-be models. Potential issues are, for example: poor communication between organizational units, too many order backlogs, or execution of unnecessary / unprofitable processes. For the evaluation of to-be processes, the same evaluation criteria should be used as for as-is analysis.[167]

Using evaluation criteria of as-is analysis for to-be analysis

The following basic heuristic principles have proven to be suitable for the generation of to-be models:

- In general, a parallel processing of partial functions within a process has to be given preference over a sequential processing, since a parallel processing will accelerate the processes. Of course, resource interdependencies must be taken into account. When two processes in a process model access an exclusive resource, then a conflict may arise on the process instance level that allows an acceleration of the processes only to a limited extent. The guiding principle reads as follows: "As much process economy as possible, as much resource economy as necessary!"[168]

Helpful Principles of to-be modeling

---

[167] See Schwickert, Fischer (1996). See Chapter 5.3.1 and Chapter 5.3.4.
[168] Horváth (1997).

Model qualification rather than company units

- A process should preferably be executed by one person or by one group. The potential savings and / or quality gains from fewer organizational interfaces in the process need to be compared with the additional costs of partially non-value-adding activities by overqualified personnel.[169] In addition, the natural tendency to incorporate organizational units into the to-be models is in fact an extremely complex task. In order to avoid the impending increase in complexity, it is recommended to abstain from including organizational units in the to-be process model. Modeling of the organizational structure and related complexity effects are moved to a later modeling phase. This method is a proven practice and reduces complexity considerably. [170]
- For individual process steps or process sections, self-control should be made possible. A later quality assurance tends to lead to a higher number of interfaces and to a decrease in motivation due to truncated competencies.
- For every process, and not only for business processes with customer contacts, a "customer" needs to be designated, if applicable, as an internal customer. This strengthens the customer perception. In addition, it facilitates a comparison between rendered services and market performances.
- When creating to-be models, it must be kept in mind that the employees who will eventually execute the processes will gradually increase their productivity. The structure of the processes should contribute to a strengthening of the employees' awareness of value-adding services. This can be reached, for example, by a higher transparency of tasks and by the awareness of the employees' own contribution to the total performance. The employees will realize the effect of their own actions and will be motivated to improve the entire process rather than just local functions. The created process models must be edited in such a way that every employee can recognize his own contribution to the company's success.

Division depth

In addition to the proven principles, other process improvement considerations need to be discussed. An important consideration is the hierarchy and the related structuring of processes on multiple levels. During the course of modeling, decisions need to be made regarding which functions of a process model justify further refinement into individual sub-processes, and to present these as individual process models in their own right. Of course, a compro-

---

[169] See Hammer, Champy (1993), p. 51.
[170] For further refinement of this procedure, see Chapter 7.2.

mise between economy and completeness has to be found. The following criteria will help to decide on the depth of refinement:[171]

- For processes to be properly coordinated, a function which shares its input and output data with other functions, needs to be modeled at the most refined level of the function hierarchy at which this interaction occurs. When, for example, within the function "Invoice Order" a function "Post Order" is executed, and the resulting document is also used by another function, the function "Post Order" has to be modeled as its own independent function, and the function "Invoice Order" must be placed higher in the hierarchy. Functions in which autonomous activities with self-coordination are executed should not be further refined.
- The processes have to take into account the diverse requirements of different processes. It is also required to make abstractions from the single process object if the processes are to be valid for a number of objects. The hierarchy must end on the level on which all potential process objects are described by a common process structure.
- The assessment of to-be processes requires a hierarchy level on which the targets of to-be modeling, such as cost and time, can be judged in their relative importance for the total process. This must be determined in view of the targeted purposes of use[172]. If, for example, a particular function of a process creates 90 percent of the total costs, then a further division of the function needs to be considered in order to enable a judgment of efficiency.[173]

Another consideration of process design is the so-called generation of variants. The modeling teams can use two different methods to arrange process structures in a hierarchy. The first method is to generate process variants very early on, i.e. already in the models on a high level of abstraction. The second method is to generate variants on the lower process hierarchy levels. Process variants basically represent the same process but differ in detailed process structures because of the description of these processes for different process objects. An example here would be the processing of urgent orders and normal orders.

Generation of variants

The primary advantage of an early generation of variants is the reduction in model complexity. Since the created single variants can be regarded as widely independent of each other, the interdependencies can be ignored in further modeling. Another advantage

Early generation of variants

---

[171] See Gaitanides (1983), p. 80.
[172] See Chapter 6.2.1.
[173] See in addition to this Chapter 3.4.

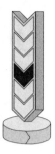

is the easy readability and, therefore, better understanding of the models, since other process models can point directly to individual process variants. The individual variants can be explicitly recognized and created with minimum redundancy. However, the model handling is clearly more complex and time consuming. This is explained by the fact that the individual variants have similar structures and, partially, the same functions. It should be pointed out that insufficient support of model administration by modeling tools still exists at this point. Here, advanced modeling tools can contribute considerably to model quality by ensuring model consistency. The decisive factor is the advantage of being able to view the individual variants independent of each other. At the same time, the disadvantage exists of overlooking possible synergy effects and utilizing less optimized resources. It is just this weakness that later offers the advantage of generating variants which then lead to uniform processes down to the last possible level of hierarchy. A later generation of variants, however, goes in line with a compulsory omission of individual process features. This can only be partially compensated by a better documentation of the processes. If the individual processes are changed while drafting the total process model, then they can be traced later with more ease to generate variants, since the relationship between the processes of different variants is lost only on the lowest model level.

*Late
generation
of variants*

*Variant
generation
at DeTe
Immobilien*

In the beginning, DeTe Immobilien selected a late generation of variants within to-be modeling. The processes on the upper levels of abstraction were not differentiated by process objects. An example here is the handling of planning processes. This handling was not distinguished by objects to be planned. This caused a uniform process that enabled a "one-face-to-the-customer" principle. This in turn guaranteed the customers a standardized order processing within DeTe Immobilien and uniform planning processes for personnel, technical, or commercial services. Similarly, uniform interfaces could be fixed in some places for support processes, such as material delivery or coordination of service staff.

*Use of
process
scenarios*

In the further course of the project, the clear necessity to customize the to-be processes arose. This included the extension of the to-be models by process object-specific attributes and structures. The employees of DeTe Immobilien, who would ultimately use the to-be models as the basis for a related organizational conversion, required detailed models. In order to meet these requirements, so-called scenarios were developed. The input in these scenarios was the company's generally valid to-be models called "generic processes". For example, the planning process for different products of DeTe Immobilien was detailed similarly to the op-

erating management or energy management. In order to avoid the total number of processes in these scenarios to be mapped with partially redundant data, specific process scenarios were combined with the generic processes. The structure of these generic processes needed to be incorporated in the structure of the scenarios, as far as this was reasonable for the related subject. Thus, the demand for a higher degree of detailing of the user departments could be fulfilled for those who used the to-be processes more and more. At the same time it could be ensured that the scenarios were built in conformity with the reference processes.

Another aspect to be considered for to-be modeling is the distinction between to-be models and ideal models. To-be models are models that can be converted within a time frame of approximately three to six months. They serve as a basis for converting the project goals and / or purposes without being modified. In contrast, ideal models describe a desired status that can be abstracted from marginal conditions and be regarded as a long-term target.

To-be models vs. ideal models

In order to make this difference clear, we take IT services as an example for an influencing factor. In an ideal model, a resource- and process-efficient process is modeled based on an integrated IT-support of the individual functions within a process. The operational reality of IT services, however, often still contains monolithic, antiquated systems which cannot provide the individual functions to support the economic functions or only at unreasonably high costs. Therefore, ideal organizational process structures are reasonable but cannot be converted within three to six months from the IT point of view. The process that can actually be implemented is not always identical with the ideal process. Nevertheless, there is something to be said for designing ideal models during process modeling because this prevents, from the very beginning, a tendency to disregard views for reasonable solutions, which at the time may not be economically viable due to temporary restrictions. The way to innovative concepts is open for the expert users and is not limited by the IT infrastructure. This can clearly contribute to greater motivation on the part of team members. Basic solutions are generated faster and the loss of time and resources is minimal.

An ideal model can be regarded as a long-term objective. It serves as a basic solution for a to-be process to be converted and as an essential basis for continuous process management. The economic concepts used to create ideal processes clearly have a longer lifetime than technical or organizational restrictions. After ideal models have been created, they should influence the design of the to-be models. If, for example, technical innovations become avail-

Ideal model as long-term objective

able and are relevant for a converted to-be process, then the originally designed ideal process can be referenced and its conversion can then be re-considered. An example here would be a delivery note that is sent via Electronic Data Interchange (EDI) as part of order processing activities. When the delivery note is taken as a basis for planning, this can accelerate the reception and inspection of goods received. If EDI is not available as a short-term option, the to-be process, in contrast to the ideal process, has to be extended in a way that the inspected goods received are compared with the order. This requires the order to be placed next to the goods received in the process.

6.2.4
Documentation of to-be models

Documentation of process features

The created to-be models must define at least the goal and purpose of the process, independent of the achieved degree of detailing. The model must also contain a detailed description of the object characterized by the process (e.g. customer order or invoice). A related name for the to-be process must be derived from the defining attributes above. The name has to be meaningful in order to make its contents clear and understandable for external persons.

The created to-be models can be further enriched with the following information, depending on the degree of detailing:

- Process models:
    - Process owner (name, organizational unit)
    - Contents and type of process change
    - Contacts to external business partners
    - Planned frequency of execution (per day, week, month, etc.)
    - Planned processing time
    - Planned costs
- Data models (also technical term models):
    - Data manager (name, organizational unit)
    - Estimated impact of information system
      i.e., how will the data be used in information systems
    - List of changed technical terms and / or data definitions
    - List of new technical terms and / or data definitions
- Function models:
    - Changed relations in the functional hierarchy
    - Planned frequency of execution (per day, week, month, etc.)
    - Planned processing time
    - Degree of (actual and future) support by Enterprise System

Often, to-be process structures are evaluated according to the experiences of the method experts, and, therefore, are based on pure quality aspects. For example: A decrease in the number of process interfaces or an increase in the number of employees in order to cut the processing times does not always lead to the desired improvement. In this case, quantitative criteria are required. However, these analyses are problematic. An improvement in a single process metric, such as decreasing costs by reducing personnel, may cause other metrics to deteriorate, such as processing time. Because of the complex efficiency relationships, the effects of single measures cannot always be anticipated. Furthermore, the quantitative evaluation of processes with modified structures imposes a great demand on the method experts. The introduction of new functions, for which parameters are not yet available, complicates this problem further. Therefore, a determination of the quantitative characteristics for processes based on process models is almost impossible.

Problematic evaluation of to-be models

6.2.5
The benefits of process simulation

To evaluate the to-be process structures, it is useful to calculate the parameters for the execution of a process, such as average processing time or average costs. This data can serve, for example, to plan the capacity or to evaluate the process with changed source parameters such as receipt of orders.[174] The calculation of those parameters based on process structure models, however, is not a trivial task. Parameters can only be measured ex-post, i.e. after the processes have been executed, using so-called process instances, or can only be calculated analytically with considerable cost. The calculation of the average processing time with linear process structures without conflicting resources is still relatively simple through summation, but the calculation of the same parameter with recurring process structures is much more complicated and often can no longer be done analytically if comprehensive models with interdependent resources are involved.

Parameters

Process simulation is suitable for viewing process instances and to evaluate the structure of processes ex-ante, i.e. prior to finalizing a model. For this purpose, the behavior of many process instances is simulated. The collected information can then be utilized to create the viewed parameters to be used to analyze the process structures. When executing the simulation runs, random and deterministic influences can be considered as well. Based on the analysis of a sufficient number of process instances, the process

Viewing process instances

---

[174] Viewing of precise personnel capacities and costs is only possible after the organizational chart is designed, and the company units and planning units are determined (see Chapter 7).

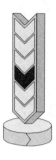

*Software
support*

structures can be modified, and the effects on the viewed parameters can be seen after repeated simulations.[175]

Simulations can be executed relatively quickly. Simulations contribute greatly to finding process structures. A proven practice to avoid the occurrence of faulty models is a validity check of the created simulation model by simulating as-is process structures with actual current data. This enables a validation, including all relevant conflicting resources. Nevertheless, the results of a simulation study have to be considered critically.[176]

*Problem:
availability
of data*

A simulation imposes high demands on the detailing of the process model being investigated. All decision rules have to be explicitly entered into the model as well as the individual functions in order to calculate processing times, capacities, or failure rates. After all related attributes are maintained in the model, a process simulation can, for example, provide insights into the optimized capacity of a Call-Center, even considering vacations or daytime-dependent capacities. In practice, the availability of relevant data represents a problem. It can only be calculated manually at a high cost.

Animation

In addition to the process simulation, modern software tools for process modeling also offer a so-called "animation" of process models. The visualization of single processes informs the user about the behavior of individual process instances. This can be useful for providing a clear picture of the related to-be processes when communicating the company's to-be models to persons who are less familiar with the modeling methods.

6.2.6
Use of reference models for to-be modeling

Before using reference models in to-be modeling, it has to be verified whether or not the structure of the reference model and its contents are suitable to serve as guidance.

The arguments in favor of reference models in to-be modeling can be divided into three categories: cost minimization, revenue maximization, and risk minimization.

*Minimization
of costs*

Reference models with cost minimizing effects are models which simplify the time-consuming and cost-intensive tasks of structuring corporate functions, data, and processes on the one hand, and which simplify learning the modeling method by providing numerous examples on the other hand.[177] Further cost reductions and revenue increases can be achieved through the use of

---

[175] See v. Uthmann (1998).
[176] See Müller-Merbach (1988), pp. 450-451.
[177] See Hars (1994), p. 32.

ideas that are described in the reference models. The fact that large parts of an already quality-tested model are implemented in a company further decreases the risks that are linked with modeling. Therefore, potential error sources in mapping the desired processes can be reduced. In addition, the interactions between the different descriptive views in the reference models are already validated several times.

Minimization of risk

6.2.7
Model optimization and consolidation

The activities of model optimization and consolidation can be clearly separated from the activities of the modeling team. While the semantic correctness of the individual models has already been secured by the modeling activities of the individual teams, the next modeling step is to arrive at an entire model by integrating the single models. Here the successive integration of new modeling complexes as they are completed is preferred rather than a complete integration of all partial models. This is because the complexity of integration increases considerably with the number of models to be integrated. The intensity and duration of this integration varies substantially depending on the selected modeling method.

Integration to an entire model

The top-down method clearly facilitates the integration of single models since they have been adapted to previously designed hierarchical process structures. Therefore, the models are already designed with a default structure of the levels above them. In most cases, the upper level is the level on which the value chains are located. This facilitates a "strategy-structure-fit" that serves to measure how well the corporate goals are supported by the corporate structure.[178]

Integration with top-down method

The bottom-up method makes structuring as well as the integration of models difficult. The reconciliation of the processes when consolidating the corporate goals is an extremely complex task that bears a number of risks. The consolidation itself requires an enormous know-how of all corporate processes in order to be able to generate reasonable structures.

Integration with bottom-up method

Independent of the selected method, the integration of the results from each individual team involves further work, i.e., model quality assurance and model quality consistency, which, in part, are also designed as control functions. The model quality assurance includes the following tasks:

Model quality assurance

- Checking the observance of modeling conventions.
- Attributing the individual model elements and comparing the required degree of detail with every individual modeling view.

---

[178] See Welge, Al-Laham (1998).

Model consistency assurance

The detection of failures in the model quality can be automated, for example, by corresponding software support. Often, missing attributes can be easily identified. Unfortunately, however, the available software does not sufficiently support the observance of modeling conventions. The quality assurance of model consistency requires the following tasks to be observed:

- The process interfaces must be checked for consistency.
- The continuous use of data and / or technical terms has to be secured. Normally, no data must be entered in a function that has not been previously created by another function. Strictly speaking, this is also valid for company-external data and information; however, if the presentation is clear, an explicit modeling of the main data entry- or retrieval functions can be omitted.
- A redundancy-free modeling of all utilized modeling objects (functions, events, technical terms, etc.) in all process models must be achieved. Preference is given to modeling tools that provide the model user with transparent information about model constructs in another location in the process model, such as functions. Functions, such as "Order is registered" and "Order is entered", should also be checked to determine whether different terms for the same activity were used. The duplicate functions have to be merged into one, eliminating the other and be used as one function with multiple uses.

Finally, the improved and consolidated models must be reconciled again by the modeling teams in order to exclude an undesired change in their economic contents. It is good practice to create a version of the to-be models after termination of the modeling task, in order to offer a basis for the tasks to follow. It is also a good idea to maintain a related history with all change requests in order to ensure the traceability of the modified procedures.

Consolidation of to-be models for DeTe Immobilien

The top-down method selected by DeTe Immobilien proved to be suitable for the consolidation of the to-be models. The modeled to-be processes could first be implemented in the previously developed upper levels of the hierarchical process structure. After the pure technical consolidation, but before the contents-related consolidation and quality assurance of the individual models the following number of models and elements existed in the database:

- approx. 750 single process models which had been refined up to the fifth process hierarchy level

Quantity framework prior to consolidation

- approx. 2,600 technical terms (incl. 1,000 technical terms from as-is modeling)
- approx. 7,500 functions and 10,000 events

When analyzing the models, and when executing quality assurance, it has been found that approximately one third of all technical terms within a to-be model were not defined at all or were defined insufficiently, and that a number of language defects occurred, such as homonyms and synonyms. Since a part of the functions had not been described sufficiently, the evaluation of the model conventions had to be secured at a later time. The reasons for the defects in the to-be models had been explained by the lack of time as well as by considerable differences between expert users and method experts in the interpretation of the contents of a particular domain. These differences and the high tacit knowledge of the modeling team of the processes led to a documentation that was in parts difficult to understand for external persons. To improve the model quality and to assure model consistency, the technical term model and the process models were revised at the time of consolidation. In spite of comprehensive quality assurance efforts, the major part of the consolidation activities was spent ensuring consistency between the single processes.

The task of consistently mapping the process interfaces between the different single processes, whose relationships were not hierarchical but rather networked, turned out to be very difficult and time-consuming. The consolidation of all single processes after the to-be modeling activities was inadequate. A decision was made to split single processes into smaller process structures in order to enable a re-use of partial processes in another place within the process model. Often, questions arose from the splitting and succeeding integration of partial processes that could only be clarified by the modeling teams that, however, had already been dissolved by this point. A more gradual consolidation of single models into an evolving total model would have enabled far better communication between model consolidators and expert users.

Incremental vs. single consolidation

After the consolidation, the number of components of the total model was as follows:

- approx. 800 single process models up to the 7th process hierarchy level
- approx. 1,700 technical terms (incl. 1,000 technical terms of as-is modeling)
- approx. 7,000 functions and 9,500 events.

Model size after consolidation

Only consolidated processes could be used to develop the organizational structure. In particular, the double occurrence of objects such as technical terms and functions, had to be avoided in order to enable a reasonable evaluation. This includes for example the assignment of appropriate qualifications to particular functions as a basis for the modeling of the organizational structure.

6.3
Editing of results

*Model
user-specific
editing*

Intranet

The created models have to be edited for the model user. This requires creating related documents for every partial purpose of modeling. Employees who did not participate in the design of the models, can now view the results, provide comments, and enter them into the models prior to their actual conversion. In addition to information seminars and conventional printouts, newer media – such as Intranet – can also be used to distribute the models in the company. For this purpose, the models need to bc self-explanatory, and, if necessary, supplemented by accompanying documents. It is conceivable here to integrate the documents into the available models in the Intranet, which have been created during process modeling and which have been entered as technical terms in the process models, and to directly use these models for distribution and standardization of the document base.

Scenarios

The use of scenarios has proven to be effective when using the models as a communication means within the company. A scenario contains a complete process for a special object, i.e., a first order or a marketing concept for a special product. This makes the process models more transparent since the decision points provided in the scenario are replaced by standard values (defaults), which results in a clearly recognizable streamlined process flow.

Presentation

Experiences have shown that the total model should be distributed in the company by presentations and personal contacts in order to avoid that the results be given too little attention by the employees. The experiences have further shown the fact that if these presentations and personal contacts are omitted, and if a reorganization of the processes is announced, the model creators and the project participants experience significant problems when answering questions from employees.

6.4
Checklist

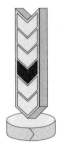

Preparation of to-be modeling

- Map the corporate strategy and / or project goals for single purposes. ☑

- Determine the modeling views to be used (data, processes, functions) as well as the degree of detailing. ☑

What to observe!

Collection and documentation of to-be models

- Identify the performance relationship between the core processes and the support processes. ☑

- Identify the interfaces between the modeling complexes. ☑

- Work out rough process structures and present them in a simple format, e.g. as value chain diagrams. ☑

- Create the to-be models for the individual modeling complexes and evaluate them based on the catalog of criteria that has been created for as-is modeling. ☑

- Distinguish between ideal model and to-be model. Take the short-term to mid-term conversion of the to-be models into account. ☑

Consolidation and editing of to-be models

- Continuously consolidate the to-be models of the individual modeling complexes into an entire process model. ☑

- Secure the syntactic and semantic quality of the process models. ☑

- Maintain the models for further use, e.g. the modeling of the organizational structure. ☑

Design of a Process-oriented Organizational Structure

Martin Kugeler, Michael Vieting

7.1
Subject and goal of a process-oriented organization

7.1.1
Process organization and organizational structure from the process-oriented point of view

The goal of a process-oriented organization is to facilitate an "optimized" execution of processes. This includes the best possible organization of business processes in terms of cost, time, and quality. This, however, is not limited to a pure sequence of functions. It also includes the distribution of tasks that determines whether or not interfaces in the organizational structure hinder the process flow, i.e., it is important to know which units execute these functions and how these units are tied into the organizational structure.

The organizational theory distinguishes between two different aspects of organization: organizational structure (organizational chart) and process organization (flowchart). The goal of the organizational structure is to divide the total tasks of an enterprise into partial work-sharing tasks, to suitably combine these partial tasks to business units, and to secure their coordination.[179] In contrast, traditional process organization is the detailed design of working processes, which deals with the sequence, duration, and range of the partial tasks that have been formerly defined in the organizational structure.[180]

Organizational structure vs. process organization

---

[179] See Kosiol (1969).
[180] See Witte (1969).

To-be processes as basis

In the traditional concept, all essential decisions – type of working organization, control systems, authorizations, and responsibilities of the units – are determined in the organizational structure phase. The succeeding process organization navigates the processes in detail through the defined structures.[181]

The process-oriented organizational design goes in the opposite direction. The determination of an organizational structure is based on the processes from the to-be modeling phase. The phase of to-be modeling determines the process structure and, therefore, defines which objects and activities will be processed, which resources, procedures, and methods will be used, and in which timely and logical sequence the tasks will have to be executed.[182]

The question of "who" is finally settled by the development of a process-oriented organizational structure – i.e., which positions and / or organizational units are built, in which form do they participate in the process tasks (authorizations and responsibilities), and how are the organizational units coordinated (control system).

The potential quality of the organizational structure depends to a great extent on the quality of to-be modeling. The selected process structure (business process framework) forms the platform for the allocation of processes and / or process tasks to the organizational units and for the assignment of authorizations and responsibilities devoid of any overlaps. The required qualifications and their timely use are an important input for the generation of positions, for their assignment to organizational units, and for their further sub-division according to the range of management activities and to the management principles of these organizational units.

Features of process-oriented organizational structure

The goal of the process-oriented organization in regard to the organizational structure does not differ in principle from other solutions. The goal is to achieve a profitable and practical organization of the company. This requires the observation of certain design criteria, such as optimized utilization of resources, a customer focus, process efficiency, motivation effects on the employees, etc.[183]

When designing an organizational structure, it is important to set priorities for the design criteria to be followed during the process of generating the organizational structure. These priorities for a process-oriented organizational structure are: minimization of interfaces, i.e., the lowest possible number of people and / or organizational units participating in process execution, uniform goals

---

[181] See Chapter 7.4.1.
[182] See Chapter 6.
[183] See Chapter 7.2.1.

and uniform criteria of success, as well as unique responsibilities within a business process.[184]

The results of this design are compared with other design criteria to find the optimum support for this business process. The design must, therefore, be adjusted, if necessary.

Let's, for example, assume that very specific legal know-how is required to execute some particular tasks of a complete sales process in a sales department. As a result of this special requirement, a legal unit must be created which, perhaps, may not be fully occupied by these tasks. In this case, a bundling of all legal tasks into a special organizational unit can become useful. Thus, in this case it seems reasonable to tolerate an interface for the purpose of economies of scale.

The important point in this development process is that it has a focus on the entire value chain. This is the main feature of process-oriented organization. Crucial to the design is not the optimization of single process tasks with different design criteria, but how the single process tasks contribute to the optimization of the total process.

Focus on total process

When the criterion "minimization of interfaces in the organizational structure" is further detailed, it means that comprehensive task structures on the unit level (job enlargement, job enrichment) are given preference over specialization. This means that an entire business process is given priority when assigned to one organizational unit at the time the organizational units are generated. If applicable, a detailed description of individual positions can be omitted. Instead, all tasks and requirements are delegated to one team that must handle a process in its entirety.[185]

The number of design priorities to be applied depends largely on the general conditions of the company. Included among these conditions are: corporate culture, environmental culture, market situations, general legal conditions, the existing organization, and the employees, as well as by the modeled processes. This results in high demands on the know-how of the people involved. This includes qualified basic knowledge of the organization, i.e., labor law, collective bargaining law, as well as knowledge of developed processes and of the daily operation of the business.

General conditions of reorganization

Furthermore, the organizational structure plays an important role in operational practice because the organizational structure shows the distribution of task contents, which aligns with reputation and salary, and, on the management level, with position and power. This can cause resistance to change and leads to influences

Rank of organizational structure

---

[184] See Krickl (1994), p. 28.
[185] See Theuvsen (1996), pp. 68-69.

in favor of personal advantages. Consequently, it is very important to clearly stress the underlying design principles.

7.1.2
Interfaces in the organizational structure – the most important joints

Most often, business processes are valid for all organizational units. Their execution depends largely on the interfaces in the organizational structure, i.e., the interfaces between the organizational units. The consequences of those interfaces are, for example:

Consequences of interfaces to organizational structure

- Extension of processing times because of interrupted material flows and information flows
- Incfficiencies among the participating organizational units due to different goal criteria and success criteria
- Increased information costs to handle and forward the process object (e.g. sales order)

Only after a detailed examination of the timely and logical relations of the process do the number and effects of interfaces in the organizational structure become obvious. From the business process point of view, the reciprocal actions from the organizational structure and process organization must, therefore, be considered and realized in order to be able to optimize the timely and logical process relations.[186]

Types of interfaces to organizational structure

When designing an organizational structure, three groups of interfaces can be distinguished.

First, interfaces exist between two positions within one and the same organizational unit. These interfaces are characterized by the transition of a process object from one position to the other, i.e., they cause an information- and / or material flow.

Second, an interface can also exist between two positions in two different organizational units. Here, as well, the transition of a process object occurs. In addition, these positions report to different supervisors because both positions are controlled by different managers.

As a result of different reporting structures, a third variant may result. In this case a task is handled by one single position and no transition of a process object occurs at this interface.

Interfaces from multiple management relations

The fact that employees are available only for a limited period of working time causes conflicts in cases where multiple management relations are concerned. Normally, division managers assign priorities and execution deadlines with their working instructions,

---

[186] See Gaitanides (1983), p. 53.

which may conflict with the priorities and execution deadlines of other managers.

These conflicts need to be completely eliminated by mutual agreement, which ties up considerable manpower resources on the one hand, and causes a stagnation of task execution on the other hand. In addition, the department managers are no longer able to reasonably manage the tasks in their fields of responsibility.

Two relevant management relations are distinguished, i.e., disciplinary and divisional management.[187] Disciplinary management is concerned with the standards of good manners and behavior. The following tasks are assigned to disciplinary managers:

- Short-term control of employees:
  - Absence-/punctuality control
  - Regulation of absence and vacation times
  - Internal approval procedures (business trips, etc.)
  - Support of employees in case of operational problems
- Long-term development of employees:
  - Hiring of suitable employees
  - Education and training
  - Appraisal / salary assignment
  - Promotion / dismissal

Disciplinary management

Divisional management refers to the modality of the executed tasks and is concerned with the following actions:

- Tasks (object / execution)
- Use of materials (type / scope)
- Information (what to consider)
- Employees (who has to do what)
- Time (from when / to when / how long)
- Site (where / where to / where from)
- Quantity (how many / how often)
- Working procedures

Divisional management

Multiple management relations for an employee must be avoided unless a temporary project organization is involved. In this case, the employee can also be under the direction of the divisional management of the project manager.

In practice, divisional management is further divided into two components: divisional instruction and guideline competence. Divisional instruction includes the type and time of the task to be executed and the personnel resources and materials to be used, while guideline competence results in the instructions for working procedures (see Figure 7.1).

Components of divisional management

---

[187] See Schulte-Zurhausen (1995), pp. 137-138.

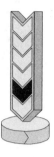

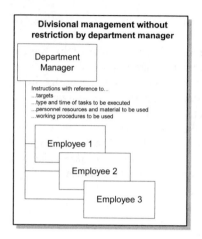

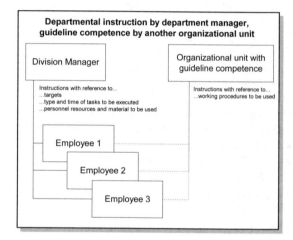

Fig. 7.1. Organizational assignment of guideline competence

Guideline competence

Guideline competence is a special case. No direct participation in the operative tasks is required. Thus, no conflicts arise from the fulfillment of the tasks and, consequently, no interface-related problems occur.

When establishing and approving procedures and methods, any conflicts have to be clarified up front in order to minimize inefficiencies in the operative execution of the tasks. Also, precise interface definitions are mandatory here. The distribution of tasks for the organizational units or positions dealing with super-ordinate tasks, and the operative positions and / or organizational units must be clearly limited.

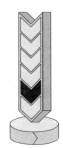

7.2
Comparison of organization types

7.2.1
Efficiency criteria of organization

The organization of a company must be goal-oriented, i.e., all organizational actions need to follow the higher-level goals of the company. It is difficult to measure the direct effects of organizational actions with respect to the company's goals. For this reason, sub-goals are derived in order to reduce the complexity of the planning problem. These sub-goals are closely linked with the company's goals. Sub-goals valid for organization are efficiency criteria. These efficiency criteria are used to evaluate the organizational actions.[188]

Efficiency criteria as evaluation scale

The two main components of efficiency criteria are efficiency of motivation and efficiency of coordination.[189]

The employees of a company should have a certain freedom concerning what should be done in terms of the company's goals. In order to achieve this, a related incentive system needs to be created. The effect of specific actions on employees' motivation is based on hypotheses and is controversial. The examination of motivation efficiency dominates the consideration of human relations and / or human resources.[190]

Efficiency of motivation

Actions to promote motivation are:
- More self-responsibility for employees, and
- Delimited and clear tasks.

The delegation of decision responsibility is intended to utilize the employees' potential creativity and detailed knowledge. The personal responsibility of the employee for his tasks is strengthened by delegation. His motivation for constructive problem-solving and fulfillment of tasks is increased and has a positive effect on the company.

Delegation of decision responsibility

In addition, clearly defined tasks have a positive effect on the efficiency of motivation since few internal performance relations exist between the different organizational units. Clearly defined tasks make complex tasks more transparent since few dependencies on other functional areas need to be observed. This enables a better collection and assignment of results that can be taken as a

Clearly limited tasks and transparent task complexes

---

[188] See Frese (1995), p. 284.
[189] See Frese (1995), p. 292.
[190] See Schreyögg (1996), p. 211.

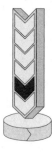

Autonomy vs. reconciliation costs

Efficiency of coordination

Aspects of efficiency of coordination

Efficiency of the market

Efficiency of the process

basis upon which to build an effective incentive system with intrinsic- and extrinsic motivation effects[191].

In addition to the efficiency of motivation, the efficiency of coordination is the second important criterion to be followed for an organizational design. Work sharing means that the tasks need to be coordinated according to the company's goals. From an economic point of view, two contrary cost developments will result. On the one hand, we have the cost for autonomy that represents the difference between the theoretically optimized decision, based on the total planning of all company areas, and the delegated decision of autonomous company units.[192] On the other hand, this sub-optimization of decentralized decisions can be reduced by reconciliation processes. The resulting costs are called reconciliation costs. The efficiency of coordination is the measuring element for the total costs of autonomy and reconciliation that must be optimized.[193]

Other efficiency criteria describe the efficiency of coordination in further detail. These criteria are:

- Efficiency of market
- Efficiency of process
- Efficiency of resource, and
- Efficiency of delegation

Market efficiency aims at a coordinated and well-organized behavior on the external procurement- and sales market. The efficiency of the sales market is higher the more the demands of the customers are addressed and coached in a coordinated way, and when economies of scale for certain customer groups, products or regions can be realized. On the procurement market, the identified demands must be consolidated in order to achieve larger quantities. The ordering of larger quantities results in more favorable negotiations with suppliers.

The criterion "efficiency of process" is used to measure the efficient arrangement of the timely and logical sequence of tasks (processes). This criterion is quantified by process parameters. These include process costs, processing times, and process quality.[194] An important feature of process efficiency is the number of interfaces in the organizational structure that can lead to inefficiencies. An increased processing time, among others, is the consequence of high transition times between different positions. In-

---

[191] See Steinmann, Schreyögg (1997), pp. 731-732 and the literature referenced there.

[192] For decentralized planning problems, see Adam (1996), p. 355.

[193] See Frese (1995), p. 122.

[194] See Chapter 9.

terfaces lead to redundant work that increases process costs. They also demand increased coordination efforts which often leads to a higher failure rate and inferior quality. An intelligent management of interfaces in the organizational structure can improve process efficiency because it allows a decrease in the inefficiencies of the process flow.

The use of resources (people, machines) in a company is measured by the resource efficiency. The goal is to utilize the resources in the best possible way, and to reach their maximum load. The efficiency of resources tends to deteriorate if multiple organizational units access the same resource, due to the uncoordinated autonomous planning of the organizational units.

Efficiency of resources

In an organization, the question of who will make which decisions on what level of the organizational hierarchy has to be clarified. Work sharing divides the interdependencies, i.e., the dependencies and links between the tasks of different organizational units. These interdependencies are viewed and defined on a higher level of the hierarchy while the lower levels tend to omit this view. The lower levels, instead, focus primarily on the tasks in their own fields of activity. As a result, an optimization of single tasks, without considering the interdependencies, leads to a sub-optimization in view of the company's goal. On higher levels of the hierarchy, a better information status and greater methodic know-how can be assumed. However, an economical reason that challenges the decisions on upper levels of the hierarchy is the increased communication and information costs this entails. Therefore, actions have to be performed on the level that has all the required information and competence. As a consequence, the only decisions which are not delegated, but decided on at a higher hierarchical level, are those decisions which cannot be made on a lower (lowest) level due to a lack of the required competence.

Efficiency of delegation

The efficiency criteria are summarized in Figure 7.2 below.

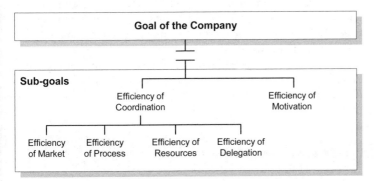

Fig. 7.2. Efficiency criteria

Between these efficiency criteria, a variety of relationships exist that can be correlated positively and negatively. Here are two examples that clarify this.

Efficiency of process vs. efficiency of resource

An often-discussed example is the goal conflict between the efficiency of process and the efficiency of resource.[195] This problem has been known for a long time in production theory under the designation "scheduling dilemma".[196] A feature of the efficiency of resource is the aim of maintaining a high degree of load on the involved resources, i.e.; a resource will be loaded to avoid idle times. A constant log of orders for which the resource is used guarantees a high and even load. If orders are always available for this resource then idle times cannot result. From the process and order point of view, this procedure is not efficient because it increases the order processing time. The orders are not processed on every workstation but wait to be processed by the responsible resource.

Efficiency of delegation vs. efficiency of motivation

Two additional efficiency criteria of mutual influence are the efficiency of delegation and the efficiency of motivation. The delegation of decisions in line with a higher scope of decision on the lower hierarchical levels normally increases the motivation of employees. However, such a delegation carries the risk that these hierarchical levels make decisions without keeping in mind the effects on other areas of the company.

As these examples show, it is not possible to follow all efficiency criteria to the same extent. A weighting of goals is required as is the case with all other planning problems that constitute more than one goal. The organizational design needs to be prioritized. This weighing has to be based on the company's strategy and goals.[197]

7.2.2
Traditional types of organization

Organizational structures can be generated based on varying criteria[198]. Different hierarchical levels can be created according to different criteria. The traditional organization theory distinguishes between two principle types of organization: Functional organization and divisional organization.

Functional organization

The functional organization is arranged into specialized functional activities on the top hierarchical level. An example of a functional organization is shown in Figure 7.3.

---

[195] See Rosemann (1996b); Mertens (1997); Reiß (1997).
[196] See Gutenberg (1983), p. 216ff.
[197] See Chapter 4.
[198] See Chapter 7.3.1.

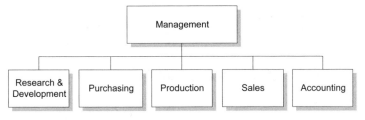

Fig. 7.3. Functional organization

Typical advantages and disadvantages of functional organizations are listed in Table 7.1.

Table 7.1. Advantages / disadvantages of functional organizations

Advantages of functional organization	Disadvantages of functional organization	
• Specialization by effective learning • Larger quantities by effective utilization of resources (synergy effects) • Delimited and controlled tasks and task complexes • Uniform appearance on the market • Tasks in conformity with traditional professional images	• Large number of interfaces that hinder the process flow (high coordination costs and friction losses) • Insufficient total view in the functional (user) departments; i.e., only their own division is optimized; effects on other participants in the process flow are ignored • Low motivation by monotonous work and missing reference to the purpose of their work • Low flexibility	*Advantages / disadvantages of functional organization*

The divisional organization[199] has a specialization characterized by objects on the top hierarchy level. Objects here can be products, regional markets or customers. Figure 7.4 shows an example of a divisional organization. Often, the generation of divisions goes in line with the delegation of responsibility for the outcome. So the divisions acquire greater autonomy, which in turn lowers the cost of coordination.

Divisional organization

---

[199] Further designations are "Business domain organization", "Branch organization" and "object-oriented organization".

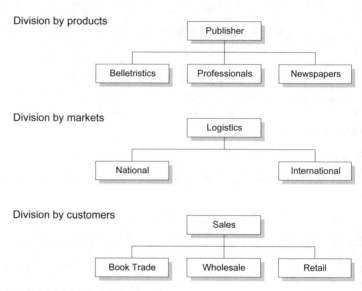

Division by products

Division by markets

Division by customers

Fig. 7.4. Divisional organization

Table 7.2 lists the advantages and disadvantages of divisional organizations.

Table 7.2. Advantages / disadvantages of divisional organizations

Advantages / disadvantages of divisional organization

Advantages of divisional organization	Disadvantages of divisional organization
• Special orientation toward relevant markets (increase in market- and customer focus) • Fewer interfaces that hinder the process flow (less inefficiencies and less coordination costs) • Increased motivation through greater autonomy • Simple calculation of success • Higher flexibility by smaller independent organizational units • Uncomplicated addition and removal of divisions	• "Cannibalism" by competition between the divisions • Efficiency losses through low splitting options of resources (synergy losses) • Aim at divisional goals instead of company goals

7.2.3
Process-oriented organization – path or goal?

The term "process-oriented organization" is often used excessively in discussions. In theory and practice, one talks about process orientation without considering this term. In most cases, process orientation serves as a pseudo argument to justify new structures, since this is a modern management concept. However, to what extent these new structures are really process-oriented is not revealed. Therefore, what is to be understood by "process-oriented organization" must first be defined.[200]

The term "organization" has two different dimensions. One dimension understands organization as a procedure; i.e., the description of rules to design a goal-oriented interaction between different people in fulfillment of work sharing tasks (= functional organization).[201] Another dimension understands organization as the result of this procedure, and, therefore, as "goal-oriented, open and social building with formal structure"[202] (= institutional organization). Process orientation refers to both dimensions of the term "organization".

The term "organization"

The procedure to design an institutional organization should explicitly include the processes in the reflections about the design actions to be taken. In doing so, processes have to be modeled in order to use them as a basis for organizational design decisions.[203] The goal of these actions and decisions sees the process-oriented organization as a building. Its structures should support the process goals (time, costs, quality) as well as possible. However, process orientation does not mean that the efficiency of resources is ignored completely. If resources are not used efficiently, it will impact both the process costs and the process goals. Therefore, the efficiency of process also depends on the efficiency of resources and, thus, cannot be regarded as an isolated item. Process orientation can only follow an altered weighting of the efficiency criteria. In organizations with strong, function-oriented structures, the efficiency of resources ranks on top. The efficiency of process is omitted. This leads to an increased number of interfaces which hinders the optimization of the process flow. The task of a process-oriented organization is to make the efficiency of process the focal point. Process orientation thus weights process efficiency higher than other efficiency criteria.

Functional and institutional dimension of process orientation

Efficiency criteria and process orientation

---

[200] For the term 'process orientation' see, in particular, Bogaschewsky, Rollberg (1998), pp. 104-105.
[201] See Krüger (1994a), p. 13.
[202] Schulte-Zurhausen (1995), p. 1.
[203] See Chapter 7.3.

Efficiency of resources and process orientation

Often, the efficiency of resources is forgotten in discussions about process orientation. Processes, however, can only follow the goal of cost minimization as one of the process goals, when the resources are utilized economically. Focusing on the efficiency of resources over the years led to a lack of attention towards the process. Process orientation, however, should not completely ignore the efficient utilization of rare resources.

Efficiency of delegation and process orientation

Another high weighting is assigned to the efficiency of delegation.[204] The dedicated delegation of decisions to executive positions can streamline and accelerate the processes since activities to collect information for decision- finding from super-ordinated jobs are omitted.

Efficiency of market and process orientation

The efficiency of market is primarily considered from the sales point of view when explicitly reflecting on the processes in an organizational design. Core processes have a customer-to-customer relationship. The orientation toward core processes during reorganization supports a customer focus as well as the efficiency of market.

Efficiency of motivation and process orientation

The efficiency of motivation is difficult to realize in practice since no empirically-proven hypotheses exist about the effect of certain organizational actions on the employees.[205] Therefore, the efficiency of motivation should be given a lower weighting than the efficiency of coordination in a process-oriented organization in order to be implemented later in the organizational process as an "adjusting screw".[206]

7.3
Modeling of the organizational structure view and its integration into the process view

A methodic support of an organizational structure design requires that the process models be extended by organizational units, etc. An isolated presentation of an organizational view is not sufficient. The process models need to show who is participating in which function and in which form.

---

[204] See Theuvsen (1996), p. 69.

[205] See Frese (1995), p. 308.

[206] See also Frese (1995), p. 312. For relations between coordination- and motivation efficiency as complementary and / or conflicting goals, see Chapter 7.2.1.

Suitable means for presenting subject-oriented concepts for an isolated organizational view are organizational charts.[207] They can be presented in different formats.[208] An example is given in Figure 7.5.

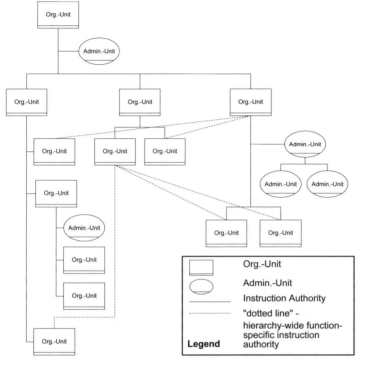

Organizational chart

Fig. 7.5. Example of an organizational chart (presentation technology)

The organizational view needs to be integrated into the ARIS-architecture in the process view[209] in order to identify the participating people, positions, or organizational units.[210] Possible symbols for people, positions, and organizational units are shown in Figure 7.6.

---

[207] See Scheer (1998c), p. 52.
[208] See Schulte-Zurhausen (1995), p. 415.
[209] See Chapter 3.
[210] The presentation of qualifications and decision authorizations only is also possible. This is done in 'role modeling'. See Chapter 7.3.3.

Symbols

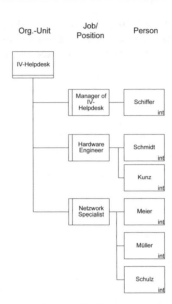

Fig. 7.6. Symbols to present the organization in process models

Integration of organizatio-nal view

People, positions, or organizational units can be assigned to functions in the process model. In particular, it is recommended that positions be mapped in the process models. Positions are independent of specific employees. Modeling individuals as part of the process models is only useful if the function to be modeled must, in fact, be executed by an actual person independent from the position (e.g. signing a contract by the managing member). Further, it is not possible to refer to actual people when a reference organization is developed which will be valid for multiple sites.

Normally, organizational units cannot be used for modeling an organizational view in process models, which is valid particularly for an organizational design. This is because they constitute a number of positions (e.g. manager of an organizational unit, secretary, and different positions for executives). The organizational design, however, needs information about which position is actually participating in a function and in which way. Since similar positions may be assigned to different organizational units, a differentiation in the names of the position is required (e.g. "buyer in regional office").

The graphical notation for the assignment of positions to functions is shown in Figure 7.7.

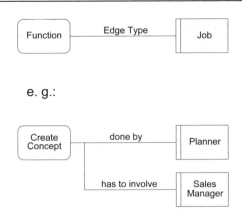

Fig. 7.7. Graphical notation of function-position-assignment

The specification attached to the connecting line between function and position reflects the type of participation of the assigned position. The following types of participation are differentiated[211] (Table 7.3):

Types of involvement

Table 7.3. Types of involvement for function-position Assignment

Type of Involvement	Meaning of Involvement Type
done by	The assigned unit is authorized to execute the function. The task is to fulfill the related function according to the quality and time requirements. If more than one person participates in the processing of the function, then their coordination lies in the responsibility of that unit which is authorized to execute the task. This includes the decision on the use of units that are identified by "can involve" (see below.). Responsibility for execution does not mean that this unit delivers the largest task-oriented input or spends more time on the execution of the function than other units. In order to secure a clear assignment of responsibility, only one execution responsible must exist for one activity.
has to involve	The assigned unit has to be consulted when processing the function. The cooperator has to be given an opportunity to deliver this appropriated support in time, and is forced to provide this supporting service.

---

[211] Other types of participation are used, for example, by Schulte-Zurhausen (1995), pp. 413-414.

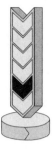

Type of Involvement	Meaning of Involvement Type
can involve	The assigned unit does not need to be consulted when processing the function. The responsible person for execution needs to decide on the participation of the unit, depending on the situation. Therefore, the person responsible for the execution of the function has a range of decisions to make when this type of assignment is concerned.
decides	The assigned unit has decision competence in fulfillment of the function. It has, however, the possibility to delegate its decision competence from case to case or in general.
is informed about	The assigned unit must be informed about the execution of the function and about the execution results.

The connecting lines can be adapted to the requirements of the company. However, the number of the different types must not be too large since this would increase the complexity considerably. In addition, a clear delimitation of the different connection lines is required in order to enable a differentiated assignment.

Mapping of guideline competence

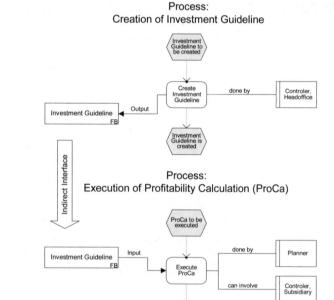

Fig. 7.8. Effects of guideline competence

The major part of the involvement of organizational entities can be mapped by assigning positions to functions. An exception is the management relation "guideline competence", since no direct participation of the organizational unit in the guideline competence exists. Instead it influences the fulfillment of the function indirectly through the guideline. In this case, the participation would be described in the documentation of the guideline as input to the function – modeled as a technical term. Additional information about this input needs to be given and described by the responsible organizational unit (see Figure 7.8).

Guideline competence

When modeling for DeTe Immobilien, no distinction was made between "has to involve" and "can involve". The only connection line in use was "involve" because the ARIS-Toolset did not allow a related "has to" and "can" differentiation. Therefore, it could not be clearly discerned from the process models which units had to participate in a certain function, and which ones only in certain cases. An example is the lawyer who has to participate only in complex contracts that cannot be covered by standard contracts. The connection line "involve" suggests that the lawyer always participates in the creation of all contracts. The fact that no distinction between "has to" and "can" was made led to unclear task assignments and was, therefore, criticized as inefficient.

Problems at DeTe Immobilien

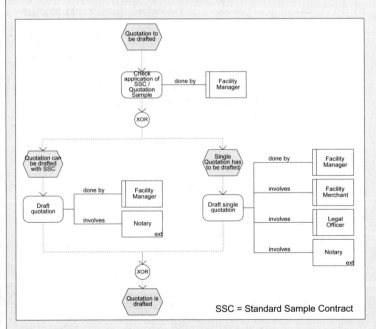

Fig. 7.9. DeTe Immobilien – Draft of quotation (example)

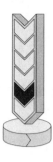

Unique assignment of responsibility

A smooth process flow is guaranteed by a clear assignment of responsibilities. This means that only one position is responsible for one function or activity. To-be modeling, however, maps generic processes, i.e.; equal process flows at different sites are modeled only once. It can be reasonable to execute one and the same process by more than one execution responsible. A divisional organization, for example, executes sales activities in every division by its own organizational units. Normally, however, the sales process is identical. This is reasonable since all sales units can then be supported by equal IT-systems and can be realized using synergy effects. For an illustration of the organization within the processes, a unique clarification of who is responsible for what and when is required.

Reference modeling and occurrences

For modeling, this means that a reference model must first be designed without the integration of the organizational view. This is done in to-be modeling.[212] The reference model then points to the individual models (i.e. actual occurrences of the process) which are integrated with the organizational view for that particular occurrence of the process. In the case of multiple models, i.e., different positions are responsible for the same process, a decision rule is implemented in the model as to when and which process needs to be run (see Figure 7.10). It is desirable to have a modeling tool which supports the maintenance of the process models. This will automatically implement changes in the reference model into the models containing organizational views, and thus, avoid inconsistencies between the models.

Mapping within complex models

An entire process does not necessarily fall under the responsibility of different positions, as may be the case with single functions within a complex model. Here, the generation of multiple process models is no longer sensible. However, the intention is to document the rule of responsibility for the related functions in the model (see Figure 7.11).

---

[212] See Chapter 6.

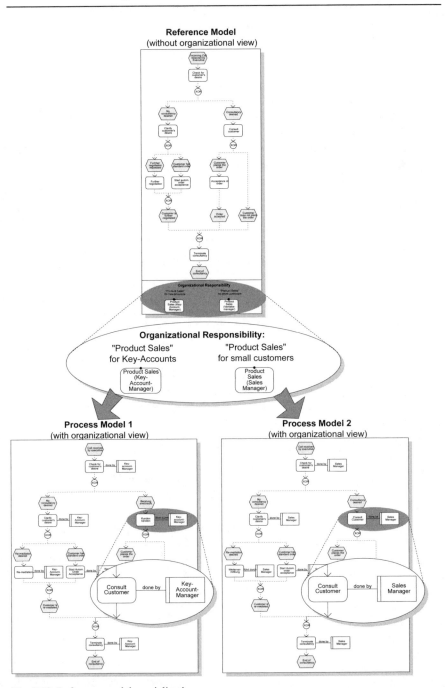

Fig. 7.10. Reference model specialization

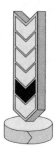

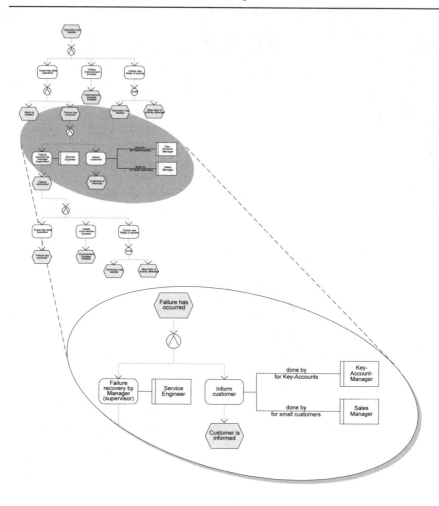

Fig. 7.11. Modeling of responsibilities within a complex process model

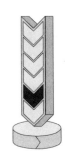

7.4
Process model to design a process-oriented organizational structure

7.4.1
Traditional method: Analysis-synthesis concept

The traditional literature on organizational theories proposes an analysis-synthesis concept as the method to develop a dedicated organizational design.[213] Here the distinction is between task analysis / synthesis and work analysis / synthesis. This leads to a 2-level analysis-synthesis concept.

Analysis-synthesis concept (2-level)

In this concept, task analysis divides the total business (tasks) of the company first into single components and then assigns these "elementary tasks" to the positions in the task synthesis. Only then can process structuring start with the detailing of elementary tasks in the work analysis and with the sequential planning in the work synthesis. Critical here is the later consideration of the processes and their design after the positions have already been defined in the organizational structure.

In contrast to this, the process-oriented organizational design focuses on processes.[214] The positions can only be created after a complete analysis of the total business of the company based on processes and, if applicable, on additional criteria. Therefore, in the following, only 1-level task analysis and -synthesis are being discussed.[215]

Analysis-synthesis concept (1-level)

Task analysis details the total business into partial tasks based on criteria down to elementary tasks. These division criteria in traditional analysis-synthesis concepts are:

Task analysis

Traditional division of task analysis

- Function (e.g. verify, enter, send)
- Object (e.g. invoice, order)
- Rank (decision- and execution tasks)
- Phases (planning, realization, control)
- Relation and purpose (partial tasks dedicated to the direct or indirect fulfillment of the main task)

The singular division after execution or after the object as proposed by the traditional analysis-synthesis concept is not far-reaching enough. The process-oriented view requires combining

Division criterion "process"

---

[213] For the following, see Kosiol (1976).
[214] See Gaitanides (1983).
[215] Gaitanides uses the term "Process Analysis" and Berg uses the term "Task Decomposition". See Gaitanides (1983), p. 61; Berg (1981), p. 68.

these two perspectives. Process is defined from object and execution and, as a result, forms an integral view of single criteria.

Process modeling is an aid for task analysis. When detailing the processes based on the business process framework, the timely and logical relations of the tasks are explicitly considered simultaneously.[216] In contrast, the mapping of a table or hierarchy neglects this sequence; for example, a divisional plan consideration of the sequence can be omitted.

Task synthesis

Task synthesis summarizes the elementary tasks from the task analysis based on certain criteria in order to be able to allocate them to a task carrier.

Traditional criteria are:

Traditional criteria of task synthesis

- Function
- Objcct
- Rank
- Phase
- Purpose of relation
- Task carrier
- Material
- Space
- Time

Criterion "customer" for task synthesis

Actual developments in many branches, from buyer to the buying market, demand a stronger customer focus. In addition to the traditional criteria, the customer must be regarded as a criterion for the task synthesis as well. A task synthesis by customers aims at more market efficiency for the buying market. The tasks for a customer and / or a customer group are assigned to a position or to an organizational unit. This results in the fact that all tasks for a customer are executed from one point (in line with the "one face to the customer principle").

Criterion "process" for task synthesis

The design of a process-oriented organization with primary focus on process efficiency requires a task synthesis according to the criterion "process". A process integrates the view of object and execution. The timely and logical relations between the tasks are mapped in the process models as well. Only this integrated view of the task synthesis enables a function-wide optimization of the total value chain.

---

[216] See details in Chapter 6.

7.4.2
Process model-aided procedure
based on a reference model

The object of the creation of an organizational structure is to distribute all the tasks required to fulfill the company's business into a work-sharing system. The procedural model presented here describes how to proceed in order to orient the design of such a work-sharing system on processes. The procedural model can be divided into several steps as shown in Figure 7.12.

In Step 1, the business process framework and the generic to-be processes must be extended by process variants. In Step 2, the organizational units must be derived from these process variants, i.e.; the execution of one or more process variants is assigned to an organizational unit.

Steps of procedural model

In Step 3, the information required for the generation of the task structures and / or positions and organizational units must be collected. The bases for this are the process variants and / or related processes, including their documented tasks. Furthermore, this requires the calculation of skills and knowledge as well as the HR (Human Resource or personnel resources) capacities that are needed to execute the tasks. Therefore, the tasks to be executed by the organizational units are defined.

In step 4, the task structures of the required personnel resources, i.e., positions are generated and assigned to organizational units. In addition, the responsibilities for these tasks are determined.

Step 5 involves the collection of the tasks that are required to guarantee the functionality of the company but have not been documented in the processes. These tasks are also assigned to positions and organizational units.

This procedure, in particular Steps 1-4, has placed the processes, process efficiency, and in line with them the minimization of interfaces to the organizational structure in the center of the organizational design.

In Step 6, additional efficiency criteria are considered, in particular the efficiency of resources. The positions and organizational units are adapted accordingly. This involves calculating the quantity of the required personnel resources and their workload capacity, and transforming the positions into established positions[217]. In addition, the quantitative consideration in this step enables the optimization of the control range of the individual organizational units.

---

[217] Established positions are vacant quantified positions. Example: The position of a "service engineer" has to be established six times, which results in six established positions "service engineer".

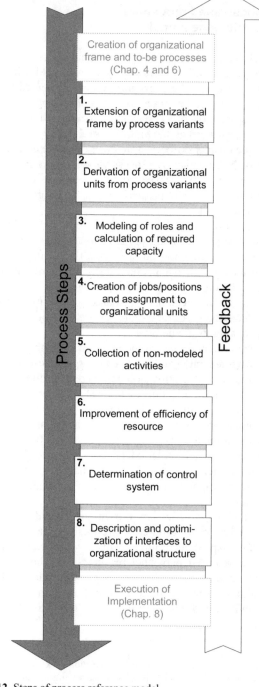

Fig. 7.12. Steps of process reference model

In Step 7, the performance system of the company is developed, i.e., the structure between management and organizational units with operative tasks. These organizational units have been derived from the preceding steps.

In Step 8, the remaining interfaces in the organizational structure are described and optimized in order to achieve a smooth-running business process, even beyond the limits set by the organizational structure.

The total procedure, however, should not be executed strictly sequentially. There are feedbacks to the preceding steps in order to repeatedly improve the defined general conditions by taking into account knowledge gained from later steps.

In to-be process modeling, generic processes are mapped out.[218] The business process framework maps equal processes without overlaps.[219] A top-down task analysis, however, may show that a differentiation of the processes according to additional criteria is useful. For example, the criterion "site" for enterprises dealing with multiple regions. In addition, strategic decisions can be implemented, such as the creation of key-account-management with the process variants "sales for key-accounts" and "sales for small customers". For this purpose, process variants are created in Step 1. Process variants are characterized by the fact that their timely and logical sequence of activities is equal since they are based on generic processes, but are executed by different positions and / or organizational units.

Step 1: Creation of process variants

In order to minimize the interfaces in the organizational structure, the tasks of a process must be assigned to one organizational unit only. First, process variants are assigned to one organizational unit (for example, process variant "key-accounts"), and the process variant for "small customers" to another organizational unit. The assignment of process variants to organizational units is the platform for further development of the organizational structure. This first assignment of process variants to organizational units, however, must be completed by a detailed consideration of every individual process task.

Step 2: Process-oriented creation of organization-al units

The timely and logical sequence of tasks is previously calculated in to-be process modeling. Therefore, the tasks need to be taken from the models of the process variants. To-be modeling does not determine the required skills and knowledge, or the personnel resources for this task.[220] This information has to be found in role modeling, described next in step 3.

---

[218] See Chapter 6.
[219] See Chapter 4.
[220] See Chapter 6.

Step 3:
Role modeling
and capacity
calculation

The documentation of skills, knowledge, and manpower depends on the defined goals in the reorganization. A minimization of interfaces in the processes leads to a comprehensive task structure for positions. Normally, a change in the assignment of tasks to organizational units and positions, results in a need to generate new positions. The process-oriented generation of positions is based also on the timely and logical task structure derived from the processes.

When determining the necessary skills and knowledge to execute a process, attention must be given to the fact that the same skills and knowledge are required to serve later as a reasonable basis for the bundling of process-oriented, comprehensive task structures. However, a reference to existing job descriptions and their names must be avoided because they already represent a combination of skills and knowledge, which often feature a functional orientation of the tasks and are not necessarily characterized by a process-orientation. The creation of a quotation, for example, may require a combination of technical planning know-how and knowledge of calculation and law. Here, it can be very useful in regard to a comprehensive task structure to further educate a design engineer in the subjects of calculation and law in order to avoid interfaces to commercial and legal positions and / or organizational units. In addition, the avoidance of job descriptions allows for flexibility in filling the vacancy. Example: For technical planning activities, the qualifications of a master craftsman would be perfectly sufficient for some products, while the technical planning for other products would require the qualifications of an engineer. In addition, the use of job descriptions may render the workload of a position difficult to manage. If the workload from the tasks in the relevant process is only 20 % for a "lawyer", then either additional tasks outside this process have to be assigned to this "lawyer" position, or this "lawyer" position has to be assigned to another organizational unit. In both cases, additional interfaces in the organizational structure would result. The alternative of an additional legal education for design engineers can make this interface superfluous as far as basic knowledge is concerned, and no special knowledge is needed.

"Roles" to
help generate
process-
oriented jobs

The documentation of knowledge and skills is facilitated by the created "roles". In the example above, the roles in the creation of a quotation could be named "expert for technical planning", "expert for calculation" and "expert for law of contract".

Which actual roles are generated depends on the related tasks in the processes and can, therefore, be designated as company-specific. Roles serve as modules for the later creation of positions. The creation of roles is based on required skills and knowledge. Normally, their focus is function-oriented ("expert for calculation", "expert for technical planning"). Keeping the module characteristics in mind, the role should not be defined in too compre-

hensive a format. Taking the quotation above as an example, the role of "expert for quotations" would constitute too many skills and knowledge, which would be better suited for one position. When creating positions, i.e., a combination of roles, the more functional roles become a bundle of process-oriented roles.

When using such roles it must be observed that all processes aim at a standardization of the required roles in order to be able to identify similar skills and knowledge in positions that are created later. Without standardization, roles which had the same meaning when they were modeled, but which later cause difficulties in recognizing them as equal roles, may result, such as "experts for calculation" or "expert for cost accounting". *Standardization of roles*

The assignment of roles to process tasks necessitates the required capacity to be calculated simultaneously. When generating new task structures, their workload has to be considered, i.e., working hours per month or man-days per year. The information and procedures used to calculate the capacity depend on the related general conditions of the company. Information about the predicted future development of the order volume is necessary. The demand of the processes and / or tasks has to be compared with the available resources, i.e. working hours, shift calendar etc. In addition, the effect of new processes and structures on the productivity has to be estimated. A precise calculation of the required resources by simulation, for example, is rather time-consuming. Therefore, it is often more sensible to estimate the required resources by experience. In any event, important sources of information are the expert users from the operative business areas.[221] *Calculation of required capacity*

The minimization of interfaces in the organizational structure means, in addition, the establishment of comprehensive task structures of the executing position within the organizational unit. Consequently, all tasks of a process – as well as the related personnel resources (knowledge and skills, capacity) – must have been identified in order to follow the goal of process efficiency. *Step 4: Process oriented generation of jobs*

The roles have to be reasonably combined into positions to form a comprehensive task structure. These positions must reflect the knowledge and skill requirements of the future jobholders in com-

---

[221] It is important to note that not every role needs the same time for the execution of the function to which it is assigned. Example: Function "Create Quotation" typically requires 10 hours. The participating roles are "Sales Expert" and "Planning Expert". From this information it cannot be derived that both, the "Sales Expert" and the "Planning Expert", need 10 hours each for this function. Instead, the share of individual work time (spent by each person) and common work time (spent by both persons in common) to execute this function must be calculated. The result could be, for example: "Sales Expert" needs 2 hours and "Planning Expert" needs 9 hours, since they work 1 hour in common to execute this function.

parison with the actual profiles of the employees and with the availability of these qualifications in the market.

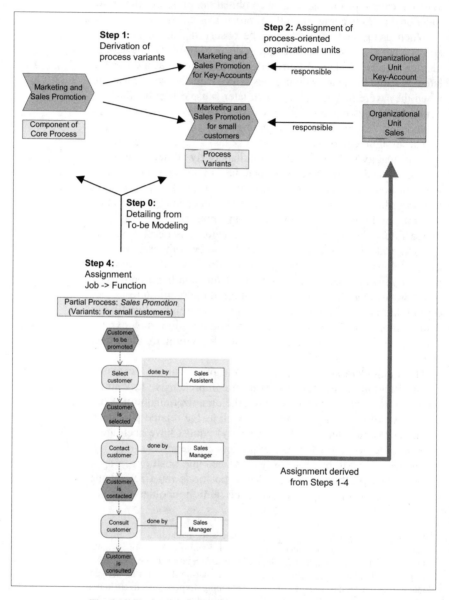

Fig. 7.13. Derived assignment of positions to organizational units

It is important to separate Step 3 and Step 4 in this procedure. The roles to be created will later serve to generate objective positions. In particular, the outcome of discussions with expert users who orientated themselves more toward old structures and a known job description, make it necessary to proceed in two steps. Often, positions are associated with actual people in the company. This adds a strongly subjective component to the discussion. If, however, the required qualification and knowledge for the execution of a function is determined, then the generation and assignment of new positions in the next step will become obvious to all participants.

Even after the creation of positions, often multiple positions participate in one task. The type of participation, however, differs considerably.[222] Some positions are responsible for the task, others are always involved, and again others are consulted only upon request, etc. An interface built by two positions, whose participation in a function is absolutely necessary, has to be ranked higher than an interface where the positions "can" be involved, i.e., they are only used from case to case.

Type of participation as scale for interfaces

At the end of this step, the generated positions need to be assigned to the organizational units that have been derived from the process variants in Step 2. Refer to Figure 7.13 for explanation.

After finalizing Steps 1-4, the organizational units and positions that have been exclusively generated to minimize the interfaces in the organizational structure are established. The mission of the following steps is to find and assign tasks which are not contained in the processes, to include additional efficiency criteria in the considerations, to design the control system, and to optimize the remaining interfaces in the organizational structure.

The tasks to be documented in business processes, as well as personnel resource capacities, are at all times only a subset of the tasks and personnel resource capacities that must be fulfilled in a company. In the first place, this arises from the fact that the primary process goals focused on the modeling of processes or process groups that were classified as particularly important.

Step 5: Tasks of company vs. tasks of processes

On the other hand, the reason may also be found in the characteristics of the process. Pure and highly creative management tasks are often not documented in process models. In addition, no support and coordination tasks – such as secretaries or receptionist – exist in models that are relevant for the processes. Tasks exist for monitoring or for legal supervisor functions, such as management staff, revision, and auditing. Also, most often, to-be modeling rejects the entry of poorly-structured processes, such as the provision of principles and methods for operative activities by buying guidelines, development of technical production processes, or by the

---

[222] For types of participation, see Chapter 7.3.

execution of annual settlements, which, however, must not be neglected when calculating the capacity.

It must be determined how far process-accompanying tasks need to be taken into account by those parts of the company that are involved in reorganization. These tasks, as well as the required knowledge and skills, need to be determined and the required capacity must be calculated.

These process-accompanying tasks must first be assigned to positions, and / or new positions must be created. These positions then need to be assigned to organizational units if new positions are concerned, and / or additional organizational units need to be generated (i.e., centralized areas such as consolidation, controlling, typist pool, etc.), which, normally, cannot be or must not be generated for process orientation.

Step 6: Optimization of efficiency of resources

The generation of positions and organizational units up to now focused primarily on process efficiency. Since, however, process efficiency is not the only parameter for an enterprise, other efficiency criteria[223] need to be included in the organizational design. In particular, a weighting must occur between the efficiency of process and the efficiency of resources.

The provided interfaces need to be checked as to what extent the positions are used (workload). All functions that are executed by a job or in which a job is involved must be listed. The timely workload of a job in a non-recurring fulfillment of the functions can be derived from role modeling in Step 3. This, multiplied by the execution frequency of the function within a defined period of time, results in the total workload for this position. The workload of the position is calculated from the quotient of total load within a defined period of time and from its timely availability. Aiming for a workload of 100 percent is not realistic. Moreover, the timely availability is less than the working time because idle times, disruptions in operation, additional activities (such as meetings, etc.) have to be taken into account.

Efficiency of resource vs. efficiency of process

The calculation of the workload of possible jobs alone, however, is not sufficient. When comparing the efficiency of resource with the efficiency of process, the specialization effects have to be investigated as well. The higher the frequency of a task, the faster the execution and the higher the efficiency of this task will be. The possible specialization effects need to be found and evaluated. Then, it has to be checked whether or not the advantages of this specialization compensate for the inefficiencies caused by additional interfaces in the organizational structure. In these cases it is useful to generate positions or organizational units which are function-oriented.

---

[223] See Chapter 7.2.1.

Example: Two functions are executed sequentially, i.e., "Order Entry" and "Order Arrangement". When one person executes both tasks, the time needed per function amounts to 20 minutes each. The processing time, therefore, is 40 minutes. If the function is executed by two people, a transition- and wait time arises amounting to 5 minutes. Since two people save time by executing one function only (specialization effect), they only need 15 minutes per function. This adds up to a total processing time of just 35 minutes. Although inefficiencies may occur through transition- and wait times, this is compensated by the time that is saved in processing.

There are also other reasons why single processing tasks are relocated to other organizational units; such as revision security (i.e. four-eyes-principle).

After all tasks of the company are collected and assigned to positions and organizational units, actual positions are derived from these abstract position descriptions which must then be filled by personnel. This means that the employees who are then actually working in an organizational unit are derived based on the total suitable workload. The number of employees in an organizational unit forms the basis for comparison with the desired control range. A too-low number of employees can give reason for integration in another organizational unit. A too-large number of employees may lead to a further subdivision of the organizational unit and / or to a generation of additional process variants in consideration of the process orientation rules.

Optimization of control ranges

As soon as organizational units with daily operational tasks or with process-accompanying tasks have been created, the organizational units below company management need to be implemented in a control system. The process-oriented design criterion here is the observance of uniform responsibility areas. This may lead to a consolidation of all organizational units that deal with a certain product group or with a defined region. Centralization can be useful in case of process-accompanying tasks (consolidation, controlling, etc.).

Step 7: Determination of control system

A process-oriented organization requires, in addition, the generation of structures that enable a constant improvement in existing processes and structures after reorganization.[224]

Up to now, activities have led to positions, organizational units, and to a control system which best matched the process efficiency criterion, while also taking into consideration other efficiency criteria, i.e., that the number of interfaces have been minimized. The remaining interfaces in the organizational structure can be of a dif-

---

[224] For details, see Chapter 9.

Step 8: Description and optimization of interfaces

ferent design.[225] The different methods of optimization depend on which types of interfaces in the organizational structure exist.

The smooth transition of a process object or an interface imposes certain requirements in terms of time and contents. The content requirement is a detailed description of the states and / or scope of information needed for transition of a process object. The time requirement describes the latest date by which work on the process object has to start and the dates by which it has to be forwarded or rerouted by the latest.

Example: When a quotation is first created technically and then released for calculation, this requires the related materials to be described in detail, including purchase prices and required working hours for certain people, and technical drawings to be prepared. In addition, information containing the expiry date of this quotation is needed.

The relationships to management have to be checked for all tasks in the processes. If multiple management relations exist, they must be specified in detail. In addition, in the case of a conflict, escalation procedures must be defined that have the final responsibility.

Example: When a central material management (merchandising system) is installed that provides a basic procedure for placement and purchase, inventory and maintenance of corporate-wide general contracts, this centralized material management does not directly participate in the operative buying tasks. Their procedures, however, determine how these operative buying tasks are executed. In this case, a guideline competence is concerned whose effects and contents have to be defined for the operative processes.

Documentation of organizational structure

In many companies, diagrams, job descriptions, and task specifications document the organizational structure. The structured approach of modeled processes primarily supports the job descriptions and task specifications of the organizational units. In the process models, the individual functions are collected and assigned to positions and / or organizational units. Often, related tools can automate the evaluation of the tasks to be executed for every single position and / or organizational unit. In order to avoid that every job description is overloaded with tasks, only the main tasks should be documented. This makes the job descriptions more resistant to change.

Consistent documentation of organization

In addition, functions with qualification requirements are stored in role models. This also allows for the assignment of the qualification requirements of the functions to the positions. Furthermore, this secures the consistency of the documentation of the organization that is composed of process models and job descriptions

---

[225] See Chapter 7.1.2.

and / or task specifications of the organizational units. This in turn results in a complete and reconciled organizational documentation.

7.4.3
Process model sampled by DeTe Immobilien

The reference process model in Chapter 7.4.2 is a typical "ideal" model. It resulted from conceptual considerations prior to the development of the organizational structure for DeTe Immobilien, and has been modified, last but not least, based on the practical experiences gained during project execution. Therefore, the reference model presented here differs slightly from the procedure model of DeTe Immobilien. One of the important restrictions with DeTe Immobilien was the time to generate the new organizational structure. This prescribed time amounted to nine months and included: the documentation of job descriptions and task specifications for the organizational units, reconciliation processes with management and staff committee, job occupation, and conversion of new processes. How formidable these efforts have been in these nine months became obvious only after viewing the general framework. All 10,500 employees of the company were involved. The complete organization was re-conceptualized, the documentation of the organization re-created, and approx. 300 processes re-implemented in the company.[226] Therefore, the re-created organizational structure did not separate the individual steps precisely, but allowed different steps to overlap due to time limitations. In addition, the top-down method was applied to the management level, which defined certain business areas in advance without explicit consideration of the processes. In some cases, this led later to interface problems with the process flows that partially suffered from a considerable loss of efficiency. Therefore, the consideration of the division criterion "process" is suggested as desirable for all activities in regard to the organizational design.

Time pressure main motive for diversion from reference model

  In Step 1, the generation of process variants was strongly oriented toward the strategic defaults of DeTe Immobilien. In addition, comprehensive consideration was given, in particular by the sales departments, as to how to address and tend to the different customer groups. First, a separation was made between internal and external customers. Then, a second separation was made within the external group of customers. The key-accounts of the external customer group needed to be gained and tended to by a key-account-manager.

Process variants based on strategic defaults

---

[226] For the implementation of the organizational structure and processes, see details in Chapter 8.

Role modeling based on standard roles

In Step 2, to begin with, standard roles were created from the information that had been collected to generate the task structures and positions and / or organizational units. Then the modeled to-be processes were completed with the addition of modeled roles in cooperation with the teams that had participated in modeling. The large number of these participating teams turned out to be a serious problem in terms of coordinating the generation and modification of standard roles. Different names for standard roles mapped to equal qualifications would have produced results that could not be evaluated. Then, the roles of the individual functions were assigned to processes. The total work for more than 300 process models must not be underestimated. The real advantage, however, is the possibility of discussing and documenting the qualification requirements without thinking of actual people, positions, or organizational units. A collection of the time- and quantity aspects of the functions was omitted for time reasons. The data then needed to be estimated later, prior to conversion.

Assignment matrixes for organizational units

In Step 3, first of all, process variants were assigned to the organizational units. For an accurate comparison of interfaces in the organizational structure with other organizational units, so-called "assignment matrixes" were developed (refer also to Figure 7.14) which contained the individual process tasks in table format. The activities from Step 5 were implemented, i.e., these assignment matrixes were already completed by the tasks that were not mapped in processes. In the actual case of DeTe Immobilien, only recurring and structured processes were documented.[227] Therefore, some tasks were left which had to be fulfilled in daily work, but which were not registered in the business process model. For com-

Amendment by activities not mapped in process models

parison, the job descriptions of the old organizational structure were used. This made the detection of a large number of missing tasks possible. New additional tasks were found using the brainstorming method. After the tasks were determined, the workgroup distributed these tasks to the organizational units or, where required, created new organizational units.

The assignment of process variants and partial tasks from the processes was in part determined by the top-down method in regard to the business areas to be created, as well as by former agreements of the expert users on the organizational units to be generated.

Assignment matrixes for positions

In Step 4, the positions were created, supported by assignment matrixes (see Figure 7.14). However, in the project of DeTe Immobilien the modeled roles were not consequentially converted to positions. The proposals for positions came from the expert user teams. They defined a certain scope of tasks and qualifications to

---

[227] See Chapter 6.

be assigned to these positions. The positions were compared with the processes using assignment matrixes in order to investigate how far these proposals supported the processes or forced the provision of additional interfaces, which conflict with process orientation. In this respect, a comparison was made between the efficiency of process and the efficiency of resource.

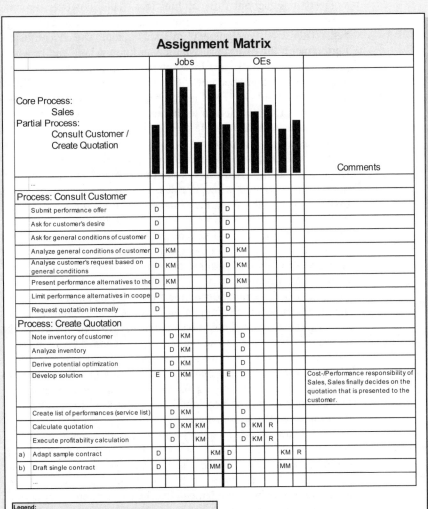

Assignment Matrix											
	Jobs				OEs						
Core Process: Sales **Partial Process:** Consult Customer / Create Quotation											Comments
...											
Process: Consult Customer											
Submit performance offer	D				D						
Ask for customer's desire	D				D						
Ask for general conditions of customer	D				D						
Analyze general conditions of customer	D	KM			D	KM					
Analyse customer's request based on general conditions	D	KM			D	KM					
Present performance alternatives to the	D	KM			D	KM					
Limit performance alternatives in coope	D				D						
Request quotation internally	D				D						
Process: Create Quotation											
Note inventory of customer		D	KM			D					
Analyze inventory		D	KM			D					
Derive potential optimization		D	KM			D					
Develop solution	E	D	KM		E	D					Cost-/Performance responsibility of Sales, Sales finally decides on the quotation that is presented to the customer.
Create list of performances (service list)		D	KM			D					
Calculate quotation		D	KM	KM		D	KM	R			
Execute profitability calculation		D		KM		D	KM	R			
a) Adapt sample contract	D			KM	D				KM	R	
b) Draft single contract	D			MM	D				MM		
...											

Legend:
D   = is done by (incl. cost- / performance responsibility)
KM = can be edited in cooperation with
MM = is edited in mandatory cooperation with
I   = result is forwarded to
E   = is decided by
R   = indirect participation of OE via guideline competence

Fig. 7.14. Extract from an assignment matrix

Advantages and disadvantages of assignment matrixes

The participation in the individual positions was marked in the matrix for every function using an abbreviation ("D": executes, "E": decides, "M": cooperates, "I": has to be informed). The advantage of an assignment matrix is the clarity of interfaces in the organizational structure. Another advantage is the possibility of quickly editing the matrix without needing to know the modeling method. However, there are also some major disadvantages. The listing suggests a sequential flow of functions. This, however, is valid only for a minimal number of processes. The links cannot be mapped. Therefore, assignment matrixes often cause confusion. Furthermore, the expert users no longer know where and in which process they actually are. In addition, the problem of redundant data occurs, which becomes still more serious when different methods are used (EPC and assignment matrix). Finally, the positions have to be transferred to the process models. On the whole, therefore, it is more sensible to document the positions and their participations directly in EPCs.

Comparison with role modeling

After the assignment matrixes were processed, they were compared with the modeled roles. Then the jobs assigned to the functions were compared with the roles. In some cases, this caused a revision of the jobs since the assigned positions did not fulfill all of the qualification requirements. Two solutions were found for this problem. Either the profile of the position had to be completed by the qualification requirement, which often forced the training of employees, or another position also had to participate in the processing of the function.

Determination of time- and quantity workframe

An actual determination of the time- and quantity metrics for the processes during the project run was omitted. Therefore, no actual need for personnel could be analytically derived in Step 6. The expert users had to estimate the number of planned positions from their knowledge of the tasks to be performed in every position. This estimated number of planned positions was taken as a basis to calculate the control range. This required minor adaptations since DeTe Immobilien used benchmarks to orient itself toward the control ranges of other companies.

Determination of control system

The already-divided business areas and the determination of tasks implicitly decided the control system (centralization and decentralization). This limited the freedom of design to generate process-oriented structures in Step 7. In addition, DeTe Immobilien did not plan to fundamentally change the control system of the matrix organization. The old organization was improved in this step by the unequivocal assignment of responsibilities. For example, the responsibilities for turnover, costs, and / or outcome were specifically assigned to positions which were also of major profit for the processes themselves, since disputes about competencies were widely eliminated.

The optimization of the remaining interfaces (Step 8) in regard to timely and logical process object- and information transitions was already well prepared through to-be modeling. The documented input / output relations of functions clearly and unambiguously defined which output had to come from which positions in order to enable another positions to take this output as input to execute other functions. This defined the transition from one positions to another positions in the process.

Optimization of interfaces based on process models

In the development phase of the organizational structure, however, management relations have not been sufficiently considered. In particular, the guideline competencies caused problems when implementing the processes. The competence for costs and outcome and / or the process competence were developed from discussions that often did not touch on all aspects. Here, a systematic analysis of management relations would have done a better job.

Problem: insufficient consideration of management relations

The documentation of the organizational structure was very well supported by this structured approach. The derivation of tasks and qualifications from process models for the description of positions and organizational units decreased the costs and the time needed to create the complete documentation. Taking over the terms from the process models enabled a consistent system for the documentation of the organization (processes and organizational structure). This uniform terminology allowed the tasks in the job description to be easily and quickly assigned to the processes. This transparency was further increased by the provision of information in the Intranet. A direct connection existed between the modeled positions in the processes and the related job descriptions. The employees could use a browser to navigate through the process models and view the job descriptions of interesting positions by double-clicking.

Take-over of terms from models for job descriptions

Problems also resulted from the fact that DeTe Immobilien uses a procedure that bases the monetary appraisal (salary payment) of the positions on the term used in the job description. Example: The term "coordination" has a higher rank than the term "support". This aspect was not considered in the to-be modeling of the processes. The terms used in partial tasks were not selected taking into consideration a later appraisal. Therefore, a direct transfer of terms from the processes to the job descriptions was often impossible.

Problem: Job appraisal based on terms

In the business-related workgroups, this know-how about the monetary appraisal of positions was absent as well. The terminology criteria were unknown to either the members of the staff committee or to the management, which later caused enormous efforts to re-work this terminology. This later work could have been avoided if representatives from the HR department, with their know-how in the appraisal of positions, would have been available earlier.

Problem: Teams lack job appraisal knowledge

In many areas, the resulting organizational structure can be regarded as process-oriented. The ideas about the organizational design focused mainly on processes. The outcome was an organizational structure where the responsibilities for the processes and / or partial processes were clearly assigned and precisely defined (responsibility for results, turnover, or costs). Competence disputes and inefficiencies could be diminished. Frequent points of discussion were the interfaces that were regarded as necessary. Interfaces should become the linking elements between positions and organizational units, and should disrupt the process flows as little as possible.

Restrictions of process-oriented organizational design

A methodical processing and conversion of a process-oriented organization is not free from problems. The processes are only criteria that influence the organizational structure. Many restrictions, such as the interests of the employees represented by the staff committee, considerations of power and influence, or existing IT-systems, force the acceptance of many compromises in the design of a process-oriented organization, since the new organization will not be a theoretical structure, but a conversion to the new orientation of the company. Therefore, the goal is to find an organization that represents a solution that is as "good" as possible in light of these restrictions and in light of the different concepts expected from this goal.

7.5
Checklist

Preparation

What to observe!

- Leave the teams from the to-be modeling phase unchanged, if possible. Utilize the detailed knowledge gained from process modeling. ☑

- Provide the teams with knowledge of labor law and collective bargaining law as well as with HR development know-how. Positions are created and appraised. They must, however, also be filled by employees. ☑

- Involve the representatives of the most important decision-makers. The results of the complex organization construction must be understood and accepted by management as well as by the staff committee. ☑

- Issue incentives. Much is demanded from the employees. ☑

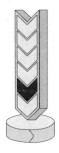

Development of organizational structure

- Proceed systematically. First find the requirements and qualifications before distributing tasks to the positions. ☑

- Focus on process and possible interfaces when thinking about the generation of positions. But do not forget the efficiency of resources. ☑

- Do not underestimate restrictions (policies for power and influence, existing IT-systems, interests of employees), which complicate a consequential process orientation. ☑

- Define and describe the remaining interfaces in the process in detail. It is not sufficient to be aware of the interfaces. Instead the cooperation across the interfaces must actually work. ☑

Documentation

- Map the organizational view in the process models. ☑

- Create a uniform documentation of the organization by transferring the terminology from the processes to the descriptions of jobs and organizations. ☑

Afterwards

- Development is just one aspect; conversion is everything. Invest in an effective change management. ☑

- Establish an organization that allows conversion to continuous process management. ☑

Process Implementation - Process Roll-out

Michael Laske, Redmer Luxem

8.1
Position of implementation in total project

An essential challenge of reorganization, in addition to the design, is the implementation of new structures.[228] For a successful implementation, the central site should avoid to develop one solution with (massive) external support that will then be sold as the (final) project result. From the design phase of a changed organizational structure to the implementation phase, the goal and the path to this goal have to be constantly reviewed and questioned. All necessary modifications should be noted, all procedural concepts should be presented, and, finally, all ultimate desired changes should be turned into reality.

There is no "king's road" for an implementation. Instead, suitable measures must be selected and reasonably combined[229] while considering the logical, political, and cultural facts of the existing organization and the scope of planned reorganization. BOURGEOIS and BRODWIN investigated the impact of different management styles on the success of implementation projects.[230] Other important factors of influence are the number of employees, the number and importance of the processes involved, the scope of changes (revolution vs. evolution), and the defined time frame.

Situation-based approach

In conclusion, it is not sufficient to apply a generally optimized procedure. Instead, situation-based procedures must be outlined.

---

[228] See Reiß (1993), p. 551.
[229] See Krüger (1994b), p. 217.
[230] See Bourgeois, Brodwin (1984), p. 241.

8.2
Roll-out strategy

Alongside the question of a suitable organizational structure to support the conversion process[231], a procedure model must be developed that indicates in which timely sequence the new processes and the related organizational structure need to be implemented.

8.2.1
Implementation sequence of organizational structure and processes

First, the principle question arises as to how to sequence the implementation of new processes and related organizational structure. Three alternative sequences are possible:

- Implementation of new processes and then adapting them to the organizational structure.
- Conversion to the new organizational structure and then rollout of processes.
- Simultaneous implementation of new organizational structure and new processes.

Organizational structure and process organization

Adopting an approach of changing either the processes or the organizational structure first may seem sensible due to the clearer scope and the reduced risk. The close interrelationship of the organizational structure with the processes, however, favors a simultaneous implementation of organizational structure and new processes. In many cases, new processes cannot be implemented at all in an environment with an unchanged organizational structure. One problem is that organizational units must now participate in the new processes that did not exist before reorganization. Another problem is that employees have to execute tasks collaboratively, but were located at different sites before the organizational structure was converted. The same is valid for the opposite situation, which means that after the organizational structure is converted. Many processes that had formerly been executed smoothly can no longer be executed without problems. In addition, there is a risk that undesired processes are implemented in the time between conversion of the organizational structure and implementation of new processes.

---

[231] See Chapter 2.3.

Based on these considerations, and in spite of the higher risk, it seems reasonable to execute the conversion of the organizational structure and the implementation of new processes at the same time.

8.2.2
Step-by-step versus big-bang

Different strategies are also offered for the rollout of processes. Figure 8.1 shows the main features of how to proceed. Related details are examined and evaluated in the description which follows.

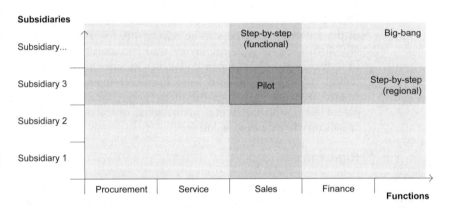

Fig. 8.1. Result after first step of roll-out

Pilot roll-out

In a conservative implementation strategy, the conversion to new processes is done in strictly one location (subsidiary) with strictly one function. This is then a pilot roll-out for additional processes to be implemented. This pilot roll-out is a regional and functional, step-by-step approach. The experiences gained here can be utilized to adapt the processes later or to refine the conversion method in other regions. Eventual problems from faulty guidelines appear only in a very limited environment and can, therefore, be better controlled and eliminated. This contributes to a very high degree of security. However, it must be accepted that the implementation time is clearly longer than that of other strategies. In addition, interface problems in strictly networked structures may arise and must be observed. Interface problems generally appear between an already "restructured" area and other areas.

Pilot roll-out

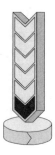

Step-by-step

Step-by-step

A step-by-step implementation successively converts the total regional or functional partial areas in the new process organization, while other areas continue to work according to the old scheme. A major advantage of this strategy, similar to the pilot procedure, is the greater safety in comparison with a big-bang. The step-by-step implementation of the new organizational structure allows the successive learning of process-related facts, both for the employees involved in these processes, and for the process organizers.

Experiences from the first implementations can then be used to make the subsequent partial implementation run smoothly. Because of the manageable number of user departments being involved in the implementation of the system, the training expenses of the process organizers can be minimized. In case of a regionally-distributed roll-out, employees who had already been trained in former steps can help their colleagues in other regions with on-the-job training. On the other hand, the increasing number of steps results in an increased number of temporary interfaces between partial areas with different organizational instructions. This may cause considerable friction losses.

Big-bang

Big-bang

A big-bang describes an implementation that takes place at the same time in all relevant locations. It creates a potentially greater profit than the step-by-step solution. The implementation time periods are shorter and no friction losses occur from organizational units working with different process versions. Area-wide processes can be converted in one step that involves all participating areas in the same way. Therefore, a newly-conceptualized process becomes productive in its totality. However, there are not only advantages but also some risks. The implementation risk of big bang is clearly greater than the step-by-step process, since the scope of the project implies greater requirements in the mastering of interdependences. For this reason, the big-bang strategy is only possible when it is based on strict project management. In addition, management must assign the project a very high priority in order to allow immediate decisions as well as rapid elimination of competing influences. This process is very demanding on the organizers because of the many organizational and IT-related risks that occur within a very short period of time. In addition, this strategy does not include a test phase where experiences could be gathered.

Evaluation of strategies

In summary, it can be said that the big-bang strategy aims primarily at a quick conversion of organization- and process improvements, while the step-by-step implementation weights the safety aspects higher. The objectives of implementation strategies have a conflicting relation to each other.

Evaluation of strategies

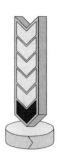

Which strategy to use depends to a great extent on the actual situation. The following table contains the essential advantages and disadvantages in a condensed format.

Table 8.1. Advantages and disadvantages of different roll-out strategies

	Variant	Feature	Advantage	Disadvantage
1	Pilot implementation	Pilot implementation in <u>one</u> region and <u>one</u> functional area	• no error repetition • very high security	• time intensive • uncoordinated anticipation
2	Step-by-step (regionally scheduled)	Implementation is done in the individual regions successively / overlapping	• gain of experience • successive optimization • creation of proliferators • high security	• high dependence on first implementation • long implementation time • synchronization need
3	Step-by-step (functional-ly schedul-ed)	Implementation is done in the individual functional areas successively / overlapping	• gain of experience • successive optimization of implementation • high security	• in workflow, only a few isolated organizational units can be observed • synchronization need between organizational units
4	Big-bang	Implementation is done simultaneously in all regions and functional areas	• fast variant • no inconsistencies in the company • all process go-live at the same time	• possibly repeated errors • difficult consolidation • high load • low security

Often, the time restrictions for the conversion to new organizational structures are very tight. This, however, is not always advantageous for a safe conversion of the processes, organizational structure, and, finally, employees when far-reaching structural changes are concerned. Because of time restrictions and in order to avoid costly, temporary interfaces to be created, the company in the presented case study selected the big-bang implementation. Here, the new organizational structures and processes were simultaneously converted in the head office as well as in all subsidiaries.

Big-bang at DeTe Immobilien

<table>
<tr><td>

Problem:
IT-systems

</td><td>

As shown, this form of conversion does not allow for all possible problems to be anticipated so that some problems occurred in several subsidiaries that a pilot roll-out would have recognized and eliminated in advance.

One problem was that the information systems were not yet completely adapted to the new processes at conversion time. After a relatively short period of unrest, the new processes were implemented in their totality. No provisionary or temporary interfaces were required, as would have been the case with a step-by-step implementation.

</td></tr>
</table>

8.3
Project marketing and information

8.3.1
Target: Acceptance

Acceptance

The conversion of the generated process- and organizational structures is often described as the most difficult part of a reorganization project.[232] In most cases, the change of an organizational structure goes in line with a change of competencies. This can be perceived by the employees as degradation. The result is, thus, a lack of acceptance by the employees, among other obstacles. Without a suitable means to prepare for acceptance at the onset as well as during the roll-out, and without massive support from the top management, these obstacles are difficult to overcome and jeopardize the success of the total reorganization. It is best to involve the employees in the reorganization process as early as possible in order to achieve their identification with the processes and to make them become "participants" in reorganization.

Large reorganization projects bear the risk that individual areas of the company have different information sources for the planned restructure. This leads to undesired speculation and to a basically negative attitude toward the reorganization project because of the fear of "uncertainties", and this will jeopardize acceptance of the total project.

Communication of goals

In order to offset such counter-productive unrest at the very beginning and in order to give all participants a feeling of security in regard to future developments, it is recommended to start with the creation and communication of the implementation concept at an early stage. A permanent and controlled communication of goals

---

[232] See Reiß (1993), p. 551.

as well as progress reports to all company areas are prerequisites for success. The offered concept serves to create clarity in all parts of the company and to secure uniform processes.

The following structure[233] of factors is a checklist that must be used when creating implementation concepts and when selecting methods to support the individual partial tasks.

Sensibility

Visions must be outlined, targets must be clearly defined, the awareness of problems must be brought to light, and the right attitude toward the project must be achieved by a living corporate culture. The aim is to provide a feeling of orientation by a clear definition of direction.

Influence

The task is to analyze the interests of participants and to identify barriers against acceptance. This way, promoters and opponents can be recognized and influenced accordingly. At the onset of the roll-out, as well as during and after implementation, the support of management must be guaranteed.

Motivation

Dedicated, convincing work promotes the acceptance of changes and motivates support for the project. The cooperation of selected employees who are involved in the solution development process causes the expert users to contribute their ideas and interests. The transfer of competencies creates a greater readiness to participate in the conversion of the innovations. This way, the related employees become participants.

Information and training

Providing information at an early stage contributes to the avoidance of unnecessary rumors. All participants (staff committee, employees, and managers, but also external participants such as customers, suppliers, creditors, and shareholders) have to be provided with the necessary information. However, just presenting a solution will not be sufficient. Instead, the participants must be convinced of the targeted improvements, of the reason why a certain process has been selected, and of the process itself.

---

[233] See Krüger (1994b), p. 213.

Training and consultancy

Taking the slogan "good service guarantees customer satisfaction" as a guideline. It is necessary to establish a hotline, to train the expert users on site, and to hold periodical meetings to support the exchange of experiences between participants.

Documentation

The clarity of processes and results is an important factor in achieving acceptance and use of new structures. This includes the conversion process (implementation documentation) and the achieved results (organization documentation) as well as the options for support (training / consultancy documentation).

8.3.2
Communication concept

Distribution of information

The draft of a process-oriented organizational structure by a central reorganization team is not sufficient to improve the processes. All members must be informed about the changes to be expected. The members may be in a position to give valuable recommendations for individual process areas that can then be used for the continuous improvement of the processes. In particular, companies with a powerful staff committee should already involve the staff committee in the project at a very early stage in order to obtain and utilize additional information channels. The attitude of the staff committee (between "support" and "rejection" of the new organization) can easily decide the success of a project. Confronting employees with an already-decided organizational structure will create a tendency to reject "the command from above" that will considerably hinder the conversion of measures.

In order to gain far-reaching information and participation of the employees involved in reorganization, it is recommended to use a number of different communication channels simultaneously.[234] The four basic elements of communication strategy are:

- Information events
- Personal discussions with employees
- Multimedia presentation of project outcome
- Publications (internal and external)

---

[234] See Schmidt (1989), p. 50.

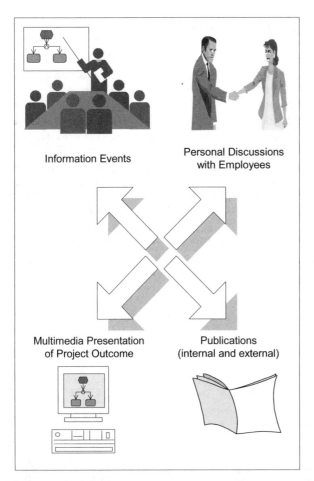

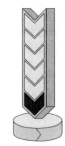

Fig. 8.2. Communication Mix

In order to convey information relating to a project's status uni- *Information*
formly with all employees, information events in all regional and *events*
functional areas of the company must be carried out. The goal of
these events is to make the employees familiar with the most important innovations. Favoring a direct information flow, it would
be desirable to have the new structures and processes explained to
every individual employee by the members of the reorganization
team who had created these structures and processes. This, however, is nearly impossible to realize for many projects in view of
the resulting workload (example: at DeTe Immobilien approx.
10,500 employees work in 12 subsidiaries all over the Federal Republic of Germany). The results of reorganization should not only

be presented, but also the effects of the detailed conversion of new processes should be discussed with the related areas.

The discussion must be initiated by the reorganization team.

Personal discussion with the employees

Details, however, can only be discussed in small groups at future conversion sites. Successful reengineering projects have shown that employees can best be reached when their direct contact partner is their immediate supervisor.[235] Therefore, the process responsibles should discuss and explain the processes with their employees on site. The basic principle suggests that individual communication can be reasonably executed only within a regional unit. The larger the company, the greater the trend to move the task of the central reorganization teams from direct explanations of new processes to the persuasive efforts of regional managers.[236] The managers must be put in position as quickly as possible to explain the new processes to their employees, and to act as proliferators (i.e. to spread communication of project information through to subsequent levels of the organization).

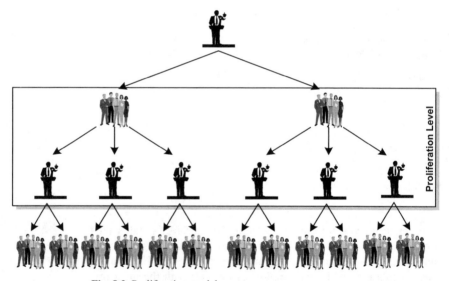

Fig. 8.3. Proliferation model

The proliferation model in Figure 8.3 divides the employees into small groups to be trained by the managers. Every proliferator, in turn, has the task of informing and training small groups of his

[235] See Lohse (1996), p. 194.
[236] See Thompson, Strickland (1995), p. 242.

own employees. This way, the processes developed in the reorganization project can be personally presented to every employee without too great a burden on the central project team.[237] In order to be able to competently answer detailed questions, the process responsibles must be supported in all professional and methodical questions by contact partners on the central site. In addition, the managers must be provided with comprehensive tools in order to make visualization of the processes easier and, therefore, assure their acceptance.

In addition to the described regional information events, a central support of the communication of project goals, as well as measures to realize these targets, are extremely important. Different media can be utilized. If available, company-internal information channels (company news, leaflets, circular letters) should be used at an early stage in order to periodically report on the reorganization status. In addition to the company-wide uniform core information, special attention should be given to providing information on specific situations in regional or functional areas, given that the company structure allows for this. The knowledge and acceptance of the project can also be promoted by external publications in appropriate magazines. Here, however, attention must be paid to ensure that the employees receive information about innovations via internal channels first. Otherwise, uncertainties and displeasure against the project management arise that may harm the success of the project.

Publications (internal and external)

A component of good project marketing should be the creation of a simple and recognizable logo. This logo should be added to all information about the project in a clearly visible format in order to facilitate an immediate association. Even if the company is blamed for poster painting, nevertheless, a catchword for the project and an easy-to-remember slogan will support a positive basic attitude toward the project.

A number of media are available to advertise the results of the project's progress within the entire company. Proven procedures such as posters and placards that constantly visualize the target of the project are suitable as modern multimedia and interactive procedures. It has proven to be useful to grant a wide range of employees access to the new processes. This access can be executed via Intranet. See detailed discussion in Chapter 8.6.

Multimedia presentation of project results

DeTe Immobilien worked out a presentation that explained a central concept of reorganization. The business process framework allowed different partial areas to be viewed and discussed in detail without losing the total overview. This presentation formed the ba-

---

[237] See Schmidt (1989), p. 51.

sis for a number of information events held by the project team in all regional units. The participants in these information events were managers who could later impart the information that was gained to their employees. The managers were also provided with overviews of special processes (ISO A0 plots) as well as to the presentation documents (through oral presentations with concrete displays or partially-animated PowerPoint shows).

In addition, regional and cross-regional editions of the company news periodically reported on the progress of planned and converted reorganization.

At the beginning of the conversion phase, the interest of the employees in the business processes was relatively low. In the early phase of roll-out, the employees reflected on quite different kinds of questions:

- How will the new company structure look?
- Will I have to move to another working site?
- Will longer commuting times result?
- Will there be a new manager?
- Will the existing job descriptions remain unchanged?
- Which new positions have to be occupied?
- Which evaluation / compensation will be provided for the new jobs?

The different assessments of jobs resulted in applications for higher-ranked positions that had not necessarily been anticipated by the project management.

Another problem was the fact that the interest in the processes was considerably less than in the innovations of the organizational structure. A further problem was the deficiency of process-oriented thinking and acting, particularly by employees of a function-oriented company, and it took some time until the employees' thinking and acting changed to a process-orientation.

After the first regional presentation event, where uncertainties could be eliminated, interest in the new processes became dominant. A problem that arose when presenting the processes was the fact that the employees normally thought in terms of quite concrete processes, whereas the (new) processes were presented in abstract (generic) format.[238] In order to take this fact into account, presentations were developed that described concrete scenarios that were then entered into the generic process models.

---

[238] See Chapter 6.2.3.

After presentation events were held everywhere, the employees asked the following questions:

- How will the new organizational structure interact?
- Which particular jobs result?
- Which IT systems have to be handled in the future?
- Which interaction results from request and feedback?

Thereafter, inquiries from the regions increased significantly and resulted in a tremendous workload, both for the central experts and for the subsidiaries. A continuous supply of information could only be secured thanks to the structured coordination of the effectively executed information events, as well as by the preparation of related presentation aids. Of particular use was the establishment of a back office to support and prepare the information events, to create the presentation documents, and to relieve the project team.

The selection of suitable proliferators represented a critical success factor. It must be pointed out that only professionally skilled people who were familiar with the daily problems of the processes to be maintained could undertake this task. Delegating this job to method experts as "professional" representatives turned out to be inefficient since the questions that were raised touched on detailed contents rather than method.

In conclusion, a very positive summary can be drawn. However, it should be pointed out that in a project of such size and duration, it must be recognized that the uncertainties of the employees can resurface again and again. Counteractions are then required in the form of dedicated project marketing.

8.3.3
Training concept

In addition to presenting "pure" information, the training of employees represents an important tool that may be a decisive factor for the success of the project. It is not sufficient to simply achieve a positive attitude toward the targets and measures of the project. Instead, all participants must be qualified to execute their future tasks with competence. This requires the planning and executing of a number of training courses prior to conversion in a new organization. The wide range of target groups, as well as the training contents, requires differentiated procedures. Table 8.2 shows the different requirements to be met by the contents of training events:

Building competencies

Table 8.2. Training requirements

	Professional	**Methodical**
Top Management	Total company ➢ overview	Understanding of value chains
Middle Management	processes for which they are responsible ➢ explanation	Understanding and explanation of value chains and EPC
Executive Level	Processes with participation ➢ practice	Understanding of EPC (Event-driven Process Chain)
Process Manager	processes to be explained ➢ overview and explanation	Modeling, understanding, and explanation of value chains and EPC

Professional and methodic training

From a professional point of view, it must be distinguished which sections of the processes need to be presented to related target groups and to what extent these groups need to be familiar with the processes and the degree of detailing. In addition, a methodical knowledge of different depth should be realized. For the "top management" target group an overview of the total processes in the company and their interaction is of major importance. Here, normally, a presentation on an abstract level will do (e. g., value chain diagrams). The representatives of middle management execute the information- and training courses for the members of the user departments and act as proliferators, as already described in the proliferator model. Therefore, middle management needs to understand and explain the methods in connection with value chains and EPCs. In addition, they must be able to give a professional explanation of all processes falling within their field of responsibility. Finally, the executives have to become familiar with the EPC models and processes in such a way that they are able to work with these processes.

Process manager

The process manager[239] has the role of a moderator who, for the most part, must fulfill two tasks: the conversion of professional defaults in the modeled process flows, and the information about the predetermined process contents, both in respect to syntactic and semantic interpretation.

Therefore, the process managers have to safely apply the modeling methods; they have to master the tools in use; and they have to be familiar with the setup of modeling standards. In addition, the understanding of the contents of processes is crucial since the

---

[239] See Chapter 9.4.3.

process managers must compare unstructured as-is processes and the requirements of the expert users with the total process model in order to guarantee the consistency of the models. The process responsibles and / or expert users are only partially able to guarantee this consistency.

The general execution of communication can be moved to the training area. Also for training courses, a proliferator model should be used in which the central project members train the middle management, and the middle management in turn trains their employees. The necessary training material, as well as prepared scenarios to use the new processes, should be provided by the central site in order to avoid double work and to guarantee a uniform procedure in all decentralized user departments.

Scenarios

8.4
Measures for personnel transfer

When implementing a process-oriented organizational structure, a number of interdependent aspects must be taken into consideration. Different time sequences for the implementation of measures can be selected.

Process models and organizational charts are available from the derivation phase of the organizational structure. For the roll-out, the assignment of the employees to the new organizational units has to be taken into account. For the assignment of jobs to employees and vice versa, the following instruments exist:
- Job transfer lists
- Task transfer lists
- Employee transfer lists

Old Job \ New Job	Job New001	Job New002	Job New003	:	:	Job Newxxx
Job Old001	✕					
Job Old002			70%			30%
Job Old003		✕				
...					50%	
...				✕		
Job Oldxxx				✕		✕

Fig. 8.4. Job transfer list

Job transfer lists

In a job transfer list (see Figure 8.4) the old and new job descriptions are shown in a matrix. The matrix fields indicate which old job is replaced by which new job. If jobs cannot be found 1:1 in the new organizational structure, this matrix can be used to enter the quantity share – but not the quality share. This quantity share must then be transferred to the new job. The share is calculated by different procedures. Possibilities are, for example, the time-share of a complete task or of a number of partial tasks to be taken over by the new job. The content of the first version (time-sharing) is rated higher, but the question arises whether or not the benefit of such a rating justifies the cost for the related data collection. The summary of rows and columns need not necessarily result in 100 percent, since it is often the case that the tasks of old jobs are either deleted or – if applicable – new tasks are added. If no time-sharing data can be collected, then the transition from an old job to a new job can be marked with a cross. This includes the possibility that an old job enters into more than one subsequent job, and vice versa.

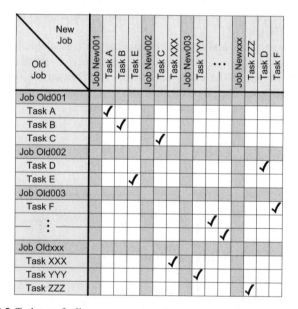

Old Job \ New Job	Job New001	Task A	Task B	Task E	Job New002	Task C	Task XXX	Job New003	Task YYY	...	Job Newxxx	Task ZZZ	Task D	Task F
Job Old001														
Task A		✓												
Task B			✓											
Task C						✓								
Job Old002														
Task D													✓	
Task E			✓											
Job Old003														
Task F														✓
⋮									✓					
										✓				
Job Oldxxx														
Task XXX						✓								
Task YYY							✓							
Task ZZZ											✓			

Fig. 8.5. Task transfer list

Task transfer lists

An effective aid in answering questions about requirement profiles is the task transfer list (see Figure 8.5). Here, the old and new tasks are not generally compared, but instead a task level is broken

down. All tasks to be executed by all old jobs are listed and faced with the tasks of the new jobs. This list assigns single tasks to the related new jobs in favor of a better overview. However, experiences in already-executed projects have shown that the costs to create such a list are hardly justified and in-depth structural changes cause this list to become so voluminous that it can barely continue to be handled. In addition, problems result from the fact that some tasks are omitted or added only after the reorganization has already been completed.

The described aids refer to an employee-independent organization. In reality, however, it must be determined, prior to the actual reorganization, which jobs need to be executed by which employees.[240] The assignment of employees to the jobs is shown in the employee transfer list (see Figure 8.6). This list indicates which employee moves from which existing job / position to which future job / position. Only one employee can always be assigned to one job, but a job can be assigned to more than one employee.[241] In addition to the necessary preparation and planning for the move, this employee transfer list can also be used to calculate the required need for education and training, and to detect gaps, if any, in the coverage of new jobs by internal employees. Therefore, the human resource department can take this employee transfer list as a basis for initiating actions for personnel procurement and personnel development.

Employee transfer lists

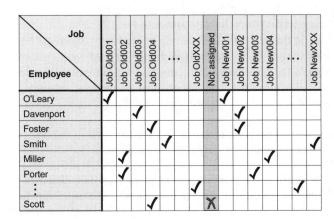

Employee \ Job	Job Old001	Job Old002	Job Old003	Job Old004	...	Job OldXXX	Not assigned	Job New001	Job New002	Job New003	Job New004	...	Job NewXXX
O'Leary	✓							✓					
Davenport		✓							✓				
Foster			✓						✓				
Smith				✓									✓
Miller	✓									✓			
Porter	✓									✓			
⋮						✓					✓		
Scott			✓				✗						

Fig. 8.6. Employee transfer list

---

[240] For procedures to assign employees to jobs, see Olfert, Steinbuch (1993), p. 108.

[241] For the (abstract) job "Service Engineer" multiple (concrete) planned jobs to be occupied by employees can be created.

<table>
<tr><td>Transfer lists at DeTe Immobilien</td><td>The staff committee of DeTe Immobilien had ordered the creation of a job transfer list. It turned out, however, that this list was not at all suitable for deriving the modified requirements of the new job owner, or could only be realized with a number of difficulties. Besides this, the large number of old jobs caused this list to become extremely long and unclear and, therefore, difficult to handle.

The conversion of the new organizational structure forced the creation of employee transfer lists. These lists had to be edited in order to clarify which employee was to occupy which new job / position. These lists were created by the conversion team in direct cooperation with the subsidiaries.

DeTe Immobilien rejected the creation of a task transfer list because the large number of tasks would have made the use of such a list uneconomical as a planning aid.</td></tr>
</table>

8.5
Technical realization

Objectives of technical realization

The objectives of the technical realization of the communication concept are to provide the users in the related user departments with a cost-effective, overall, and stable (i.e. failure-safe) process. At the time of modeling, selected expert users should be given the ability to trace the status of the modeling activities, in order to take necessary actions, if required. The following questions are relevant for the selection of realization alternatives.

- Should the available information be read-only or must it be possible to modify, if required (read-only- or read-write concept)?

Relevant considerations to select an alternative

- Should the information be accessed by all interested employees and, if applicable, also by external employees, or only by a closed user group (information market- or closed shop concept)?

- Should the existing technical infrastructure form the basis or is there a possibility of using new technologies?

Technological alternatives

For the distribution of process models via electronic channels, several alternatives exist that make distinctions in terms of functionality and technical costs. The realization by proprietary mechanisms, i.e. solutions that are exclusively dedicated to this purpose, is on the software that has already been used for process modeling. The assignment of authorization to read the database, or the installation of special read-only clients, allows a read access to the process models. In a groupware-system, i.e., in an information system for the information technological support of groupwork – e. g., Lotus

Notes or Microsoft Exchange – the employees work with a common database and can not only read this data but can also change this data (depending on the authorization). When the system is based on Internet technology, the employees can only read the data. The data to be read must be administered centrally. Also possible is a combination of each method, i.e., the transfer of groupware information based on Intranet technology.

8.5.1
Realization by proprietary methods

The ARIS-Toolset used for modeling includes the option to install read-only clients (see Figure 8.7). In contrast to complete clients that were used for modeling, these clients do not have a write option to existing process models. The fact that the users are obligated to log on to the system allows the user administration to restrict write access to visible models. Such a restriction of write access is also possible in the complete versions of the ARIS-Toolsets. The users are assigned read-only rights to the related process models in the authorization system while their authorization to write access is rejected. From the technical point of view, both solutions are equal in value; though for client licenses, the ARIS-Toolset differs in price.

Toolset or Navigator

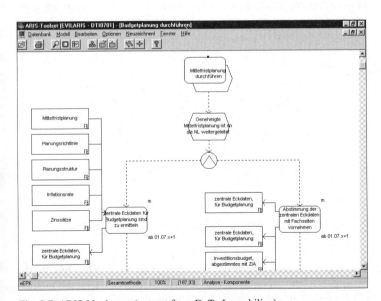

Fig. 8.7. ARIS-Navigator (extract from DeTe Immobilien)

Access during modeling

Problems of performance

The technical realization of the communication concept represented a challenge for DeTe Immobilien because of the numerous addressees of the process models. Decisive factors for the selection of technical realization alternatives formed the existing technical infrastructure as well as the internal communication policy of the company.

The decision between the installation of read-only clients or a particular assignment of rights in the ARIS-Toolset was made by DeTe Immobilien in favor of the ARIS-Navigator for economic reasons because a group license for the ARIS-Navigator already existed. This allowed selected expert users to use read-only clients to begin tracing the course of process modeling during the modeling stage, in order to forward eventual requests for correction on to the method experts. The in-time availability of information, however, turned out to be problematic since the readers were not provided with information about the actual process modeling status, via progress reports, for example. Consequently, the users complained that the process models were partially incomplete while these process models were still under development. For this reason and for performance reasons, a second (read-only) server was installed that only contained consolidated and agreed process models. This server was reconciled periodically and compared with the models of the modeling server in order to avoid inconsistencies. However, here again, it was not possible to provide the users with progress reports due to the technical restrictions of the software. The related procedure is shown in Figure 8.8.

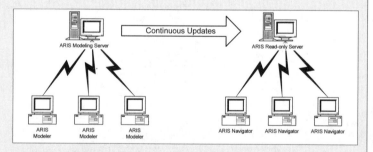

Fig. 8.8. Data transfer between the ARIS servers

Problems of user interfaces

The user interface of a read-only client that differed only slightly from that of the complete modeling client together with the technical installation costs, caused considerable handling problems, in particular, when inexperienced users were concerned. Errors also occurred from the inferior presentation of long filenames with missing 32-bit support of the Navigator. Since at modeling time no information about the handling of the client and / or the use of the modeling method was available in the company-owned Intranet, the users had a number of questions. The installation of the client turned out to be time-consuming and expensive since the existing network safety concept had to be altered for individual users. The performance of the client was insufficient in view of the numerous users so that an alternative had to be found to realize the process model in order to make it available for everyone, everywhere at any time.

Moderation or decentralized processing

8.5.2
Realization by groupware

Groupware-systems, such as Lotus Notes or Novell GroupWise, offer the possibility of providing the users with a common database. This database can be both a database with contents that have been modified by a central process owner and that cannot be changed by the end-user, and one with commonly-processed data (e.g. development documents) that is maintained as a "shared repository". Groupware-systems are client-server systems where the data is stored on one or more servers that are accessed by decentralized clients or by employees. In most cases, the installation of related client-software is a prerequisite for the use of the available information. Meanwhile, some GroupWare-systems offer the option of using the functionality of stand-alone clients via Java and / or ActiveX with standard www-browsers, such as Netscape Navigator or Microsoft Internet Explorer (see Chapter 10.6.4).

Groupware technology

The functionality of groupware-systems is not limited to access to commonly used documents, but in most cases they also allow the use of standard office functions such as a common calendar, address management, and the integration of Internet services such as e-mail. This additional functionality, however, was not necessarily decisive for the provision of process models in the company. Systems such as Lotus Notes allowed a controlled availability of redundant data (replication), i.e., the users are provided with a database in the form of multiple separate copies. The distribution of databases contributes to the avoidance of performance problems caused by remote data access.

Additional functionality

The available information in groupware-systems can be moderated by the users as well as maintained by single decentralized employees. The assignment of rights on employee- and group levels allows a detailed administration of access rights down to individual employee levels since, in any event, these employees must be recognized as users in the groupware-system. In particular, the

addition of recorded comments or sample documents for existing process models can be done more easily with groupware-systems than with comparable proprietary systems.

DeTe Immobilien did not have a groupware-system at its disposal. Since the installation of such a system was rather expensive, the value added from a decentralized handling of process models was rated too low in order to justify the additional costs that were required to install a groupware-system.

8.5.3
Realization by Intranet concepts

Intranet of DeTe Immobilien

DeTe Immobilien had an Intranet available that is based on the Internet technologies TCP/IP (Transmission Control Protocol / Internet Protocol) and HTTP (Hypertext Transfer Protocol). Every employee with a networked PC workstation can call up information from the Intranet and read it using a www-browser such as Microsoft Internet Explorer or Netscape Navigator. However, only the Intranet administrators can modify the contents. In contrast to a groupware system, a manipulation of data by the users via Intranet is not possible.

ARIS-Internet-Navigator

For this reason, this Internet technology was selected to distribute the process models in the company. The models were translated by the ARIS-Internet-Navigator from proprietary ARIS-database format in the HTML-format that can be read by the www-browsers. The process models were exported as GIF-graphs

Export of process models

that were embedded in the related HTML pages and converted via Javascript into context-sensitive image maps. For every function an individual HTML page was created that could be called up by clicking on the related item in the process model. Operation and layout of the HTML pages are oriented toward the defaults of the ARIS-Toolset. In order to integrate the process models into the company-owned Intranet, a closed domain was established in the Intranet that could only be accessed after entering a password. The page layout was based on samples (HTML-templates) that conformed to the Corporate Identity / Corporate Design of DeTe Immobilien (see Figure 8.9). In addition, information about the project (see Figure 8.10), about the processes, about contact partners, and about the method in use were stored so that they could be called up by interested employees. In particular, the enrichment of the process models by additional information lead to a positive response from some of the users.

The export of the process models into the HTML format was done on a dedicated computer since the export of the published

200 process models took up to 48 hours. Since the version of the ARIS-Internet-Navigator had no scaling option, and the process models were in some parts very large in size, the selection of the correct graphical solution required time. In addition, further technical problems in connection with the Javascript technology arose that could be eliminated only by an update of the applied www-browser version. Since the company used several versions of Microsoft's Internet Explorer, it was not possible to test every browser version with every process model in the test phase prior to model release. This, in particular, caused problems with inexperienced users. It could not be identified whether the reason for these problems (slow building-up of images, computer crashes) came from handling errors of the users or from a faulty configuration.

Design of the Intranet Site

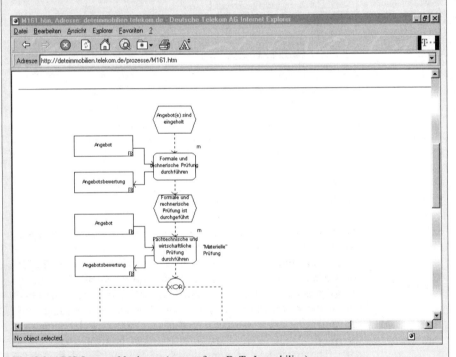

Fig. 8.9. ARIS-Internet Navigator (extract from DeTe Immobilien)

Since the conversion of the process models into a HTML format is a batch process, the conversion run can be time-controlled automatically in order to place the process models in the Intranet in time. Because of the huge data quantity that made a complete export of all models nearly impossible, this option was omitted.

Up-to-dateness of process models

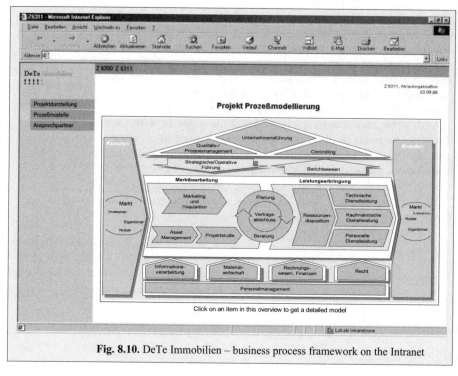

Fig. 8.10. DeTe Immobilien – business process framework on the Intranet

8.5.4
Realization by hybrid methods

Combination of Intranet and groupware

The ongoing integration of groupware- and Internet technology (e.g., in the form of a Lotus Domino-Server that converts Lotus Notes data into HTML format) makes a combination of the described methods possible. In this case, the complete client of the groupware-system will be installed for selected users. This client allows the manipulation of the available data. At the same time, the remaining users will get a read access to the available data via the World Wide Web. Therefore, the data is always kept up to date. The insurance of consistency of the generated models is achieved like the pure groupware solution through accompanying, organizational measures or the use of a centralized moderator.

8.5.5
Comparison of methods

In addition to technological and organizational efforts, as well as the possible redundancy of existing data, other decisive factors to consider when selecting a technological alternative for the realization of process models in a company are the required actuality of the available data and the existence and necessity of additional information about the type of administration. Table 8.3 compares the individual methods. Which method makes sense for the company in question must be decided for every individual case. Essential factors to consider in making the decision, however, are the existing technical infrastructure as well as the frequency in which data is changed and the target group.

Selection criteria

Table 8.3. Comparison of different alternatives.

Method	Administration	Actuality	Technical Efforts	Organizational Efforts	Data Redundancy	Additional Model Informations
Proprietary	Centralized	Very high	Client-Software	Medium	No	Depending on tool version
Groupware	Centralized/ Decentralized	High	Client-Software	High	Yes	Possible
Intranet	Centralized	High	www browser	High	Yes	Possible
Hybrid	Centralized/ Decentralized	High	www browser	Medium	Yes	Possible

8.6
Checklist

Determination of procedure

- Leave the project organization unchanged. Extend it in the regional subsidiaries.

- Determine a roll-out strategy that is suitable for your company. Thoroughly think about risks and processing speed and weigh them up.

What to observe!

Information and communication

- Identify the promoters and opponents at an early stage and focus on their individual visions. ☑

- Use different media for the distribution of information and for general project marketing! Acceptance by the employees is decisive for the success of the project. ☑

- Use dedicated proliferators. Relieve the central organizers and arrange for personal discussions with the employees. ☑

- Never forget to train all employees according to their needs (method- and user-specific training). ☑

- Pay attention to the necessary relocation of employees in time, including all related problems. ☑

Technical conversion

- Identify the information needs. Determine the access rights to data: either read-only or authorized to change. ☑

- Determine the technological and organizational responsibility for the technical conversion. ☑

- Compare your existing infrastructure with your need to present the process models. ☑

- Execute test installations with a small number of process models. ☑

Continuous Process Management

Stefan Neumann, Christian Probst, Clemens Wernsmann

9.1
From process-oriented reorganization to continuous process management

After implementation of a new organization, this organization must be managed and controlled the same as a project and needs to be adapted to changing conditions. Processes, tasks, resources, and goals for operation – the "life" – of the implemented processes are framed by continuous process management.

In addition to accompanying process implementation, the main task of Continuous Process Management (CPM) is the constant, incremental improvement of the process organization. This improvement must be in conformity with the defined corporate goals, must be based on the existing organization, and needs to include all process participants. CPM corresponds to the concept of "continuous improvement" or to the Japanese equivalent "Kaizen" but is separated from non-recurring, process-oriented reorganization methods that distinguish themselves by uniqueness and that usually possess special initiators. Continuous process management is an alternative that can be compared with Business Process Reengineering (BPR) according to HAMMER and CHAMPY. They recommend rebuilding the company quasi "on the green meadow" without consideration of existing organizational processes and structures[242]. The essential differences between these two concepts are listed in Table 9.1.

CPM versus BPR

---

[242] See Hammer, Champy (1993).

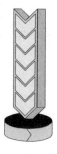

Table 9.1. Continuous Process Management vs. Business Process Reengineering (according to Bogaschewsky, Rollberg (1998), p. 250)

Continuous Process Management	Business Process Reengineering
Orientation toward existing tasks and processes	New definition of tasks and processes (understanding processes and re-engineering of processes)
Incremental, constant improvement process	Innovative, non-recurring change process
Focus on single process sections is possible	In principle, total process view
Engineering of existing organizational structures (interface management)	Non-recurring implementation of process organization (interface reduction strategy)
Consideration of all organizational goals / efficiency criteria	One-sided priority of process efficiency; resource efficiency by use of IT
Relative stability with controlled change	Instable conversion
Bottom-up procedure	Top-down procedure

The considerable differences cause the continuous process management projects and reengineering projects to be treated as real alternatives that mutually exclude each other.[243] However, a successful process-oriented reorganization that is not continued in a permanently- institutional and systematic process management seems to be unrealistic for the following reasons.

Changed environmental conditions

Adaptation to changes

The strategy of process orientation requires the ability of the company to permanently adapt to the changing conditions in a turbulent environment. In reengineering projects, the "strategic creativity"[244] is missing. Only a continuous process management can guarantee this strategic creativity.

Necessity of continuous interface management

Complex processes and interfaces

In practice, the extraordinary success that is attributed to Business Process Reengineering can only be reached in processes of little complexity. Unclear design problems cannot be handled or even mastered in short-term projects, but need to be solved through evolution.[245] This is valid in particular for the management of process

---

[243] See e.g. Emrich (1996), p. 53.
[244] Schuh, Katzy, Dresse (1995), p. 65.
[245] See Theuvsen (1996), p. 80.

variants and interfaces between the core processes that are often omitted in reengineering projects.[246]

Conflicts between different process goals

The success of reengineering processes only refers to selected sub-goals in a goal system (normally, process efficiency and improved utilization of know-how of employees when delegating decision competencies) while other sub-goals, such as resource efficiency or market efficiency, are omitted. [247] The "optimization" of a process cannot necessarily be defined by a few sub-goals, instead, "a complex mixture of goals and side conditions"[248] are concerned which must be continuously analyzed and prioritized with detailed consideration of every individual case.[249] In addition, it must be periodically controlled whether or not the costs and performance goals of the individual processes are reached. In case of deviations from the goals, or in case of changes of the process- or task structures, suitable actions must be taken to re-adapt these deviations to the related goals. The ongoing control of process conversion and periodic reviews of the modeling and planning results can result in the initiation of new improvements.[250]

Achievement of process goals

Implementation problems and barriers

Normally, the implementation of reengineered or regenerated processes is not done immediately but involves a modification of basic process models. This can be the case, for example, if the proposals for redesign and / or optimization are not detailed enough to be put directly into practice and therefore require further individualization. The (re-)engineered processes, thus, first need to be stabilized, consolidated, and further developed Continuous process management is an ongoing task that accompanies the communication and conversion of processes. The organizational entity in charge for CMP acts as a contact partner for inquiries about process methods and contents, and in particular, in the case of interface-related problems. In this respect, also during the reorganization, already-formulated improvement suggestions are put into place and adapted case by case if they cannot be converted immediately in the existing form. [251] In addition, the absence of

Accompanying conversion

---

[246] See Reiß (1997), p. 112.
[247] See Theuvsen (1996), p. 77.
[248] Mertens (1997), p. 111.
[249] See Mertens (1997), p. 111.
[250] See Heib (1998), p. 152.
[251] See Al-Ani (1996), p. 145.

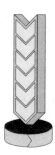

Promotion of thinking in processes

process orientation in the thoughts and actions of participating employees and managers can hinder the successful conversion of the processes. Therefore, process orientation can only be built and developed step-by-step.[252] Continuous process management is process-oriented by nature and, therefore, also plays an important role in the modification and / or change of competencies. Middle management, above all, does not only influence the process owners, but also is constantly confronted with a modified understanding of management through the delegation and decentralization of decision competencies.[253]

Implementation of process-oriented control instruments

Step-by-step implement-ation of new concepts

Process-oriented control instruments, such as process- controlling, activity-based costing or workflow management, can only be implemented in the whole company in steps and require existing and strong processes. Selection of suitable instruments and the planning, controlling, and auditing of their implementation are also tasks of process management.

The logical consequence of a process-oriented reorganization of a company is to establish a continuous process management. Process management implies the anchorage of process orientation in the total company, not limited to core processes, and systematically links the processes with still-existing central departments.[254]

The transition from a reorganization project to continuous process management can be a floating procedure. A smooth transition from the organizational structure is also possible if the participants in the reorganization are also assigned roles in continuous process management, where they can benefit from their experiences.

9.2
Goal and operation

Demands of company groups

The goal system of a company and its elements – the individual partial goals or goal components – are primarily set and modified by the main stakeholders, i.e., the capital market and the employees. The capital market expects at least a protection of investment, a (possibility of) return on investment, and a payment of reasonable interest. Often, the capital market also expects participation in the management of the company. The employees expect their working engagement to be honored by reasonable salaries and

---

[252] See Reiß (1997), p. 113.
[253] See Al-Ani (1996), p. 147.
[254] See Osterloh, Frost (1998), p. 119.

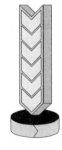

wages and to be motivated by adequate incentives.[255] These expectations are summarized by the terms: job-enrichment and job-enlargement[256].

In addition, the interests of the customers, suppliers, and of the state as well, have to be reasonably considered in the company's goals. A single, super-ordinate goal will only be found in exceptional cases. Instead, goal conflicts resulting from the different interests of different groups must be taken into account. Therefore, a primary management task is to define, to weigh, and to document partial goals.

The achievement of goals needs to be backed up by suitable actions. This requires a goal hierarchy to be built where every goal is refined by sub-goals. Every sub-goal needs to be regarded and detailed as a goal by itself. The strategic corporate goals are continuously further detailed until operative goals result that can be measured. Every operative goal must define who has to reach what, where, when and why (see goal dimensions in Table 9.2). The real goal is defined by the process ("what"); the other dimensions describe the validity range of the goal. An example of such a goal hierarchy is given in Figure 9.1.

Generation of a goal hierarchy

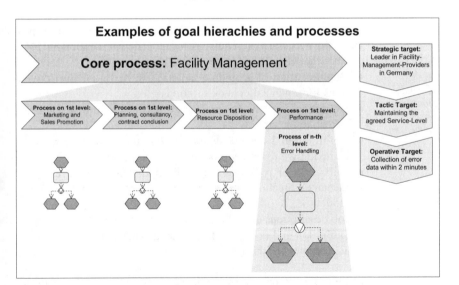

Fig. 9.1. Examples of goal hierarchies and processes

---

[255] See Hahn (1996), p. 12.

[256] Job-enrichment summarizes the goals that will increase the motivation and satisfaction of the employees by extending their decision- and control range. Job-enlargement tries to reduce the horizontal work-sharing and related monotony by increasing the diversity of tasks and goals.

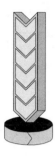

Table 9.2. Goal Dimensions

Goal Dimension	Description	Example
Who	Person responsible for the achievement of the goal	Department Manager Org / IT
What	Description of main process goal	Error rate below 5%
Where	Organizational area for which the goal applies	Western region; Process "Material Management"
When	Time period being valid for the goal	in 2001
Why	Relation between goal and super-ordinate goals of the company	Strategic corporate goal "Increase of turnover by 30%"
(How)[257]	(works with)	Use of SAP R/3 MM

Operational goals require nominal parameters to be defined for every individual process. Here it is useful to be oriented toward the "magic three", i.e., "time - cost - quality". For example, every process needs parameters to be defined for process costs, processing time, and process quality.

9.3
Phase and process model for continuous process management

Termination of the first conversion in the new organization provides the necessary prerequisites for an efficient, continuous process management. These prerequisites are:

Start situation of continuous process management

- Determination of a binding language in the modeling conventions for organizational descriptions.
- Definition and documentation of individual processes being known to the employees.
- Definition and documentation of goal and nominal parameters being known for every process.
- Definition of responsible organizational units for every (elementary) function, and related documentation in the process models.
- Job descriptions and filling of the jobs with employees.
- Definition and description of methods and tools to measure the achievement of the goals.

Ideally, the company provides its services and internal support after reorganization as described in the process models. Process instances are created and executed for every process object to be handled. Unfortunately, this status is only a theoretical assumption.

---

[257] The "how" can be pre-defined. It should, however, be left to the related employee in the sense of job-enlargement and job-enrichment.

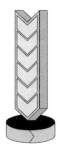

Processes require continuous management. This is especially true for a process that turns out to be ineffective or inefficient due to such circumstances as a change in the business environment caused by the introduction of new employees, by acquiring other companies, by gaining new customers, by penetrating new markets (the often-stated "globalization") or by the availability of new technologies.

Management processes are generally divided into phases that repeat periodically.[258] This is also valid for the CPM-process. Four phases can be identified: implementation, analysis, goal redefinition, and modeling. A phase model is shown in Fig. 9.2.

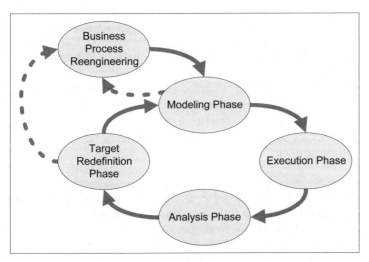

Fig. 9.2. Cycle of continuous process management

9.3.1
Implementation phase

The implementation phase has no typical start and end date. Therefore, the execution of business processes can be regarded as continuous.[259] Monitoring is an essential function during process execution.

---

[258] See Wild (1982), p. 33.

[259] Exception may result when critical errors or malfunctions are detected in a process, and when it seems reasonable to stop the execution of all instances of a process type in order to prevent damages from occurring.

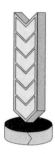

Process monitoring

If standard values (defaults), or even upper limit values, are defined for a process type (e.g., error recovery in less than three hours), the term "monitoring" summarizes all activities in connection with the continuous supervision of nominal values to be kept. In addition, monitoring also includes triggering suitable actions when the nominal values are surpassed and / or initiating related counter-actions in order to prevent such excesses from occurring. Monitoring is an effective instrument for detecting disturbances in a process within a suitable time frame. This is of special importance in cases where response times are agreed upon with the customers in order to avoid risks of great economic damage or even damage to people (response to emergency calls, for example, when an elevator has become stuck, when a fire alarm is activated, etc.).

Collection of process data

In addition, process data results from operational performance. This data is divided into two classes that are based on the data sources. The first class contains the instance data. The most important data relates to:

- Partial processes and / or process alternatives to be processed
- Dates on which the process events occur
- Jobs which execute a function
- Attributes to process objects, e.g., service level of a maintenance contract

Often, this data cannot be collected and saved because of the high cost. This must be done by information systems that support data entry and data storage fully automatically. Normally, this function is executed by workflow management systems.

The second class of data is exceptional data that only results during process execution when special events and unexpected disturbances occur which need a dedicated response. This also includes the submission of proposals to improve process designs.

In order to collect and evaluate the process data systematically, the time must be divided into (evaluation-) periods of equal lengths, i.e., days, weeks, months, or years. Every time period represents the first phase of a process management cycle. Therefore, the single process instances can be exactly assigned to one evaluation period. Normally, the assignment criteria are the start date or the end date of a process.

Involvement of staff committee

Data collection, however, is a delicate topic in view of the existing legal and social conditions. The personal rights of freedom for employees, privacy issues, and operational co-determination are often touched upon and are quickly violated. Therefore, it is recommended to involve the staff committee (union) from the very beginning and to have it participate in all decisions.

9.3.2
Analysis phase

Based on the predefined values (defaults) for to-be modeling and based on the execution data from the preceding phase, the individual processes are analyzed and evaluated to find out whether or not they match with the goals of cost, quality, and time.

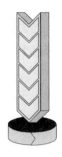

Collection of runtime data and availability

The collection of runtime data is difficult when processes run over a long period of time, when partial processes are supported by different information systems, when different jobs and people participate in process execution, and when the runtime data is available in different data formats or even only in a non-digitized form. In order to systematically edit the data for analysis, three steps are required.

First, the physical data must be made available, i.e., the data must be collected from physical storage locations. If digital data is available it has to be extracted from different systems by suitable data export functions or database queries. If the data is only available on paper, the data has to be entered manually. It must be ensured that the data is collected from all operational sites. In principle, the company must determine whether the data has to be delivered by the organizational unit where this data is prepared or whether the data has to be requested from an organizational unit (bring or carry obligation). If the data has to be delivered by the organizational unit, then this unit must be obliged to submit this data.

Utilization of data sources

In the second step, the data must have uniform syntax. In addition to a uniform data record design, uniformity is required for country-specific formats, such as a notation for date and currency. In order to define the company-specific syntax, an orientation toward existing reference models or standards is recommended. The use of standards simplifies a comparison with external sources, (e.g., participation in benchmarking studies), but also increases the flexibility in selecting additional systems to be incorporated into the data editing cycle. Modern workflow management systems often support the format that is developed by the Workflow Management Coalition (WfMC) to describe process models and process instance models. It can be expected that in the future more systems for the monitoring and controlling of processes that support the indicated standard formats will be available.

Data editing

Finally, the semantics of the collected data have to be edited. Normally, a process which passes through many different information systems and organizational units is registered under different

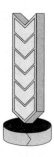

numbers (identification codes). "Sales" works with the customer order number, "purchasing" works with the purchasing request number, and "disposition" works with the internal order number. Therefore, the individual process elements must be assigned to the related super-ordinate process in order to get an error-free and gap-less history of the process runs.

Process evaluation

The collected runtime data can be compared with the default values. The succeeding error analysis classifies the detected errors by type and source. To separate the system errors from random errors, quality regulation cards, for example, can be used.

Reasons for not achieving targeted nominal values may be found in:[260]

- Insufficient description of interfaces to preceding and succeeding processes
- Insufficient support by information technology, working documents, and operating material
- Too complicated process design; the employees cannot become familiar with the process models
- Insufficient mapping of different process objects which force the generation of additional process variants
- Functions being assigned to wrong jobs / positions
- Unclear tasks
- Unclear regulation of how to cope with exceptional situations

9.3.3
Goal redefinition phase

A result of the analysis phase may be a statement that the relevant environmental conditions have changed since the time when the goals were defined. In any case, the validity of the defined goals has to be constantly verified and adapted to the new recognitions.[261] If the goals have changed fundamentally it is sensible, in most cases, to leave the process of continuous process improvement and to change to a comprehensive Business Process Reengineering.

---

[260] See Theuvsen (1996), p. 70.
[261] For description and goal definition see Chapter 9.2.

9.3.4
Modeling and implementation phase

After the goal definition phase, the error analysis must be examined for changes that have to be made explicit in adapting to the defined goals. Primarily, the process structure needs to be changed. The scope of change can vary considerably. Smaller changes include the completion of required documents or the correction of inaccurate job assignments in the process model. Large changes force the establishment of one or more specialized project teams who re-model large sections of the process or even a number of processes.

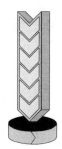

Immediate need for change

A need for change may also result for areas that support core processes, such as personnel management, information technology (IT) and asset management.

As experiences have shown, an effective and efficient IT-support of the processes is a critical success factor because – in spite of the increasing familiarity with information technology – many companies are still taxed by an adequate integration of both, processes and IT, and between individual IT-components. Companies seldom succeed in defining and converting the correct depth of integration. Only a few companies find the right way between a poorly integrated, heterogeneous and inefficient IT-support and projects that are expected to produce "solutions for all and everything - everywhere" while ignoring their inherent complexity.

When the need for change is known, the following actions must be initiated:

- A pre-study needs to be performed and / or a project needs to be set up, depending on the estimated scope of changes.
- Any weaknesses need to be eliminated by suitable optimization measures.[262]
- The responsible process manager orders the operative changes in the process model from the process organizers.
- The changes have to be communicated in the company.[263]

At this point, the cycle of continuous management is closed. The implemented changes are "implemented" and data to judge the efficiency of the changes must be collected during process execution.

Another result of the modeling and implementation phase may be a completely new concept for large parts of the process. Then the cycle is exited, and a BPR (Business Process Reengineering) project is executed.

---

[262] See Chapter 6.
[263] See Chapter 8.

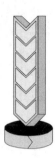

9.4
Institutional anchorage of process responsibility

Process management, in an institutional sense, is the "sum of people who are responsible for a process"[264]. In the process-oriented sense, the organization is structured horizontally in contrast to traditional function- or department-oriented responsibility concepts.[265]

Process and line responsibility

This will not necessarily eliminate line-oriented structures. They still can exist in parallel with a process-oriented division. In single cases, the authorization for decisions or instructions is interpreted by the participants in a different way. Line responsibility includes the management and the continuous improvement of the organizational unit that executes the process step in different processes. The focal points are activities with a direct effect (change, convert, adapt, settle, etc.). Line responsibility can be anchored centrally or decentrally. Process responsibility can also be anchored centrally or decentrally, but includes the management and improvement of processes whose activities mostly have an indirect effect (observe, compare, develop, propose, etc.). The features of line- and process responsibility are shown in Table 9.3.

Table 9.3. Line- versus process responsibility

Line-responsibility	Process responsibility
Process-wide view within an organizational unit.	Area-wide view within a process.
Definition and supervision of line-oriented goals.	Definition and supervision of process-oriented goals.
Development and direct conversion of problem solutions in predetermined processes.	Development of solutions when quality goals are risked; finding of potential improvements.
Delivery of inputs for process improvement.	Creation of proposals for process improvement.
Conversion of process improvements and disposition of personnel.	No direct access to organizational units in order to adapt them to the optimized process.

Task carriers in CPM

In order to realize a consequent and continuous process management, suitable roles must be defined with firmly assigned responsibilities. The distribution of tasks in continuous process management resembles that of reengineering, and the reengineering teams can be easily divided into groups with related, continuous tasks. The key drivers in a continuous process management are process responsibles, process owners and process managers.

---

[264] Franz (1995), p. 119.
[265] See Franz (1995), p. 119.

9.4.1
Process responsible

The process responsibles are of central importance for reorganization and continuous process management. In contrast to reengineering projects, the process responsibility in continuous process management is more a secondary task.[266] The process responsible provides for an efficient and effective execution of the process, for the continuous evaluation of process goals, and for all process improvements from the start of the process up to the end of the process.[267] Therefore, he has to be provided with sufficient competencies. His main tasks can be described as "information, communication, coordination and control, and permanent improvement"[268]. His instruments are monitoring, and time- and capacity control, as well as management of information systems.[269]

Process responsibility as continuous task

In principle, the process responsibility can be anchored independently with an existing line organization or with an original process organization without causing discrepancies between line- and process responsibility. When the process organization deviates from the primary organizational structure, the process responsible has to make sure that the organizational units do not exclusively follow their own interests.[270] In addition, the process responsible represents the process members when reporting to the supervisor, i.e., to the process owner. The supervisor himself "does not act as supervisor but rather as moderator or coach".[271].

DeTe Immobilien subdivides the process responsibility into centralized and decentralized responsibility. The central process responsibles are responsible for the professional execution of their partial processes in the whole company and must secure these partial processes in order to be up to date. Normally, the process responsibles are managers on the second and third management level who were designated by the process owners. The process responsibles in the central office are responsible for the following tasks:

Process responsible in central office

- Company-wide definition of process goals
- Support of process owner in his responsibility for the complete core- and / or support process
- Determination of improvements of higher priority
- Issuing orders to execute improvements

[266] See Reiß (1994), p. 13f.
[267] See Schwarzer, Krcmar (1995), p. 45.
[268] Strohmayr, Schwarzmaier (1995), p. 267.
[269] See Scheer (1998b), p. 76.
[270] See Striening (1988), p. 164.
[271] Osterloh, Frost (1998), p. 135.

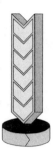

- Issuing orders for the process optimization of central projects
- Company-wide release of process changes and their execution

Process responsibles in subsidiaries

In addition, process responsibles exist in each of the 12 DeTe Immobilien subsidiaries. A process responsible in a subsidiary is designated by the related manager of this subsidiary. He details the company-wide process goals for the subsidiary; he releases subsidiary-specific process variants in agreement with the central process responsible; and he initiates their execution.

Process improvements of medium priority are determined, prioritized, and ordered by the process responsible of the subsidiary in cooperation with the line-responsible.

9.4.2
Process owner

Distinction between Process responsible and process owner

The literature about process management often ignores the process owner or equates the process owner with the process responsible. The term "process responsible" is the German version of the English term "process owner". In the case of a complex process with a hierarchical structure where the business processes on the top level are refined by processes and partial processes, it is recommended that the process responsibility be distributed to process owners and process responsibles, depending on the process level.

The process owner is situated on the top management level. The process owner is responsible for the achievement of the process goals, similar to the process responsible. He defines the process goals and matches them to the corporate goals. As a member of management, he cannot execute all the tasks of a process responsible himself, and, therefore, as a supervisor, delegates the responsibility for partial processes to process responsibles. [272]

Members of management as process owner

DeTe Immobilien basically assigns the total responsibility for the individual core and support processes to the members of management where the process owner has a higher rank than the process responsible. The process owner defines strategic goals and decides on the actions to be taken for highly prioritized process improvements in cooperation with the related process responsible and central quality manager. In the case of discrepancies in process optimization, the process owner takes ultimate responsiblity. In addition, the process owner designates the process responsibles for "his" core and support process.

---

[272] See Krahn (1998), p. 194.

9.4.3
Process manager

The tasks previously performed by the project manager and reengineering consultant in the initial reorganization project are combined and allocated as the ongoing job and / or position of the process manager throughout continuous process management. In addition, the process manager supports the process owner and process responsible in the daily management of their process-oriented activities.

The process manager's task is to coordinate all activities for process modeling and process improvement in the whole company. He collects the individual results of modeling and consolidates them into a total process model. His integration functions allow him to initiate and moderate discussions between the process responsibles. In addition, he offers the process responsibles method support and promotion and takes qualified actions to develop process-oriented thinking.[273]

Coordination of all process-related activities by the process manager

In addition, the process manager ensures the consistency and progress of future re-engineering projects. He develops and administers a pool of procedures, instruments, and employee know-how that can be accessed in future projects.[274]

DeTe Immobilien assigns the tasks of process management to the organization department and distributes the process responsibility between the central office and subsidiaries. The process manager in the central office has a company-wide responsibility for the methods used in the business process models of the company. His knowledge of methods, the connections between all processes within the company, and best practice processes enable him to fulfill the following tasks:

Tasks of the process manager in the central office

- Company-wide responsibility for control and coordination of improvements
- Company-wide identification and security of synergies from all improvements
- Coordination of cooperation between the process responsibles
- Participation in prioritizing company-wide process improvements
- Accompanying the execution of improvements
- Coaching of process owners and process responsibles

---

[273] See Scholz, Vrohlings (1994b), p. 121.
[274] See Hammer, Champy (1995), p. 151.

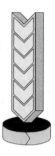

Tasks of the process manager in the subsidiary

On the subsidiary level, one full-time process manager exists in each subsidiary, whose main task is to control improvement actions. He accepts proposals for improvement and arranges for documentation and initial prioritizing of promising optimizations. He also determines subsidiary-wide measures in cooperation with the regional process responsible and quality manager. Furthermore, the regional process manager moderates the process improvement team and documents the results of workshops.

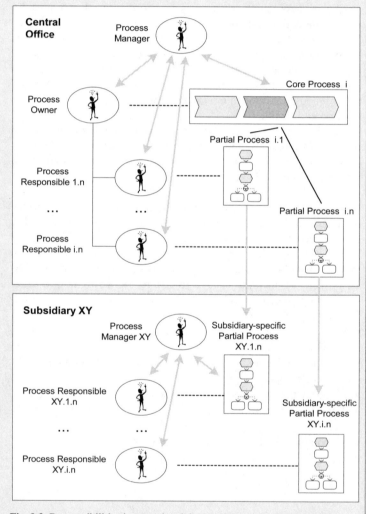

Fig. 9.3. Responsibilities in central and decentral process management

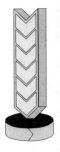

This methodic responsibility of the regional process manager covers analysis, modeling and evaluation of subsidiary-specific processes. In addition, he offers method support to process responsibles in his subsidiary and coordinates cooperation with the central office in all questions related to the business process model.

Figure 9.3 gives an overview of the responsibilities and the interaction of process owners, process responsibles, and process managers in the central office and in the subsidiaries.

9.5
Checklist

Preparation of continuous process management

- Plan a continuous process management from the very ☑ beginning to be established after the end of the reorganization.

- Define a goal hierarchy and detail the goals for every ☑ partial process.

- Make sure the staff committee agrees with your con- ☑ tinuous process management.

Management cycle for continuous process improvement

- Monitor the achievement of goals periodically and ☑ evaluate the processes.

- Use the results of this process evaluation to adapt the ☑ goal system and to optimize the process structure or process-supporting domains.

What to observe!

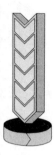

Institutionalization of continuous process management

- Designate continuous process responsibles for all processes even if the line organization remains. ☑

- Institutionalize the process responsibility also at the top management level by designating process owners. ☑

- Ensure the coordination and consistency of all current process-related activities by process managers in the organization department. ☑

- Execute a continuous process management in a regionally distributed company at the local sites as well, by utilizing regional process responsibles and managers. ☑

Additional Application Areas and Further Perspectives – Beyond Re-engineering

Michael zur Mühlen

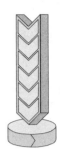

10.1
Certification

In order to ensure long-term survival in the market, the quality of a product is of extraordinary importance, in addition to cost and delivery time. Certification is one way to demonstrate to customers that the processes and structures of a company are suited to produce high-quality products. The German institute DIN (Deutsches Institut für Normung) and the international organization ISO (International Standardization Organization) have issued the standard series DIN ISO 9000-9004 that can be used by any branch of a company receiving this certification. These standards define the requirements of a quality assurance system (QA-system), which involves organizational structure, responsibilities, procedures, processes, and the required tools to carry out quality management. *Quality assurance system* The establishment of comprehensive quality management and the certification of the QA-system result in a potential increase in profit for a company. In the first place, an expectation to identify rationalization potentials is created, that in the end results in cost *Benefits of certification* reduction.[275] Secondly, by presenting the QA-system to the customer, it is documented that the company has the ability to produce high-quality products and / or to render high-quality services, which may have a positive influence on the buying decision of the customers.

The most important document for the creation and introduction *Quality assurance manual* of a QA-system is the quality assurance manual (QA-manual).[276] The QA-manual documents a company's QA-system. It is a con-

---

[275] See DGQ (1991), p. 18.
[276] For the creation of QA-manual see DGQ (1991).

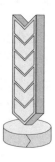

Procedural, working instructions and test prescriptions

tinuous reference manual for the management and maintenance of the QA-system and is used for the internal training of employees. At first glance, it documents for customers and other interest groups that the company has a suitable QA-system available. In addition to that, a QA-manual forms the basis for agreements on quality assurance measures with suppliers, agencies, etc.

A QA-manual is also an instrument for the management to communicate the cooperate objectives. Furthermore, a QA-manual documents who is responsible for quality management in a company. It is supplemented by procedural, working and test instructions that specify how and with which tools a task must be executed and which jobs exist between the working steps.[277] These documents describe procedural instructions for employees. Procedural instructions refer to a process with multiple steps, while work and test instructions refer to activities at the workstation.

Integration of processes and procedural instructions

Procedural and working instructions correspond to the flow of steps in performing a business process, specified by business process models that must be created in the to-be modeling stage.[278] It is recommended to derive process and working instructions from existing process models in order to avoid redundant documents. Figure 10.1 gives an example. In addition, the consistency of the models is augmented since it is a description of a continuous process model.

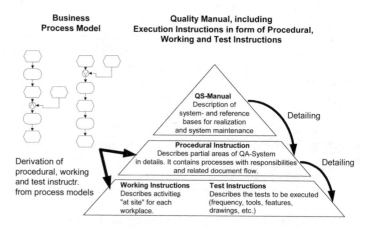

Fig. 10.1. Derivation of execution instructions from process models

---

[277] See DGQ (1991), p. 45.
[278] For to-be modeling, see chapter 6.

The derivation of procedural and / or working instructions from process models can cause some problems. The process model must be documented with many details in order to make such a derivation possible. However, this prerequisite is often not fulfilled because the processes are usually described in a format suitable for organizational design, and the functions are only documented in rough detail. Therefore, a detailing of the processes is required. In addition, elementary procedural and / or working instructions exist that do not need to be considered or completed for process modeling concepts. Further, a modeling of attributes for processes and partial processes, and / or for certification purposes, is required. For example, an attribute would be the date of certification for every individual process.

Deviation of abstraction level

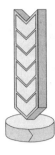

In addition to the deviating level of abstraction, the different views of the problem domain are another problem in the derivation of procedural and working instructions from process models. When designing a business structure using process modeling, other aspects than those stated in the documentation of procedural and working instructions are given priority. Process models, for example, may also contain automated functions that are irrelevant for a procedural instruction since the employee must only know how to operate the machine, and need not necessarily to know how the machine actually works. The documentation of procedural and working instructions must, therefore, contain only those functions that are operated by the employee. In contrast, information like this is of great importance for the organizational redesign in order to be able to detect potential rationalizations.

Different views of Problem domain

When designing a to-be model, an attempt was made to derive tool-supported procedural and working instructions from the documented to-be models.[279] Reports were generated from the existing models using the ARIS-Toolset. The design of these reports was based on delivered templates for ISO 9000 certification.

The completion of the QA-manual required the integration of additional documents that could only be done manually. Since the existing process models were generated on a higher level of abstraction than required for certification, the existing models had to be refined. The costs for this refinement were estimated to be higher than the costs for new procedural instructions. In addition, only selected processes were required as procedural instructions. Therefore, DeTe Immobilien considered the targeted, compact recreation of relevant procedural instructions, based on existing process models, as the more economical solution.

---

[279] For tool-supported derivation of procedural and working instructions from process models, see Helling (1998).

10.2
ERP-software

Enterprise Resource Planning (ERP)

ERP stands for Enterprise Resource Planning and is considered a continuation of MRP (Material Requirements Planning) and MRP II (Manufacturing Resource Planning). In contrast to these concepts which focus on the requirements of manufacturers, ERP is more neutral in the areas it addresses and more comprehensive in the coverage of functions since it supports, for example, the entire (internal and external) accounting or HR management. ERP-software is defined as "integrated adaptive application software" that supports both the core processes and the support processes of a company. In addition, ERP-providers offer more and more solutions for inter-company- and company-external processes such as Supply Chain Management, Customer Relationship Management or Sales Force Automation. ERP-software is often tailored to the requirements of important branches (vertical industries). This tailoring includes the elimination of irrelevant functions as well as the addition of branch-specific functions. In contrast to the established MRP and / or MRP II, ERP cannot yet be characterized by a theoretical, generally-accepted concept, but more by the common features of solutions offered by large, often worldwide acting ERP providers (mainly SAP, Oracle, PeopleSoft, J. D. Edwards).

Inter-company and company external solutions

Customizing of ERP-software

The adaptation (customizing or tailoring) of ERP-software requires comprehensive implementation concepts to be created and related IT support as an integral part of the ERP-software. This includes models that explain the activities that are relevant for system implementation and that support the execution of these activities, i.e., software for project management or pre-configured (test-)systems. Beside these tools, ERP-software is described in more detail in the form of software-specific reference models (also: reference application system models[280]).[281] Since, however, the possible (customizing-) alternatives are normally not documented in the models during system design, these reference models have an explanatory function (which processes are supported by which software and how) rather than a design function, as is the case with branch-specific reference models.

Reference models

Normally, software-specific reference models contain process-, data-, and organization models, and, to some degree, object models. The individual models are linked together in a number of dif-

---

[280] See Rosemann, Schütte (1999), p. 24.

[281] For SAP-Models see Curren, Keller (1998); Keller, Partner (1999); for Baan-Models see Van Es (1998); Kohl, Schinn (1998) and for reference models for Oracle applications see Erdmann (1998).

ferent ways. Organization models in ERP-systems describe the system organizational units that essentially structure the individual ERP-modules. Examples for such organizational units are cost accounting units, companies with balance sheet obligations, production sites, and sales channels.

A special complexity of ERP reference models results from the vast number of models (example: the SAP R/3-System is documented by more than 800 process models), so that sophisticated navigation models are needed which help to reduce this complexity by selecting single information models or information objects.

Complexity management

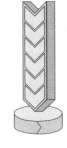

Another possibility for the management of the complexity of software-specific reference models are business process frameworks in the ERP-software. Business process frameworks serve as a starting point for navigation through the models. Figure 10.2 shows the business process framework (Business Control Model) of Baan-software for assembly-to-order producers, as an example.

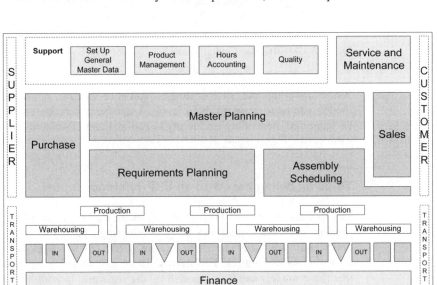

Fig. 10.2. Baan-Business-Control-Model Assembly-to-Order (source: Baan (1997))

This ERP-specific business process framework forms the entry point in finer modules from which it is possible to branch into the concrete function-, process-, and organization models. Figure 10.3 gives an example of how the Purchase-module from Figure 10.2 can be structured.

ERP-specific business process framework

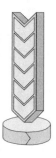

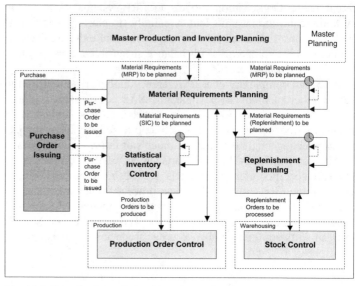

Fig. 10.3. Baan-Business-Control-Model Purchase (Source: Baan (1997))

When adapting the ERP-software, software-specific reference models can be used on different levels. Three basic procedures can be identified, depending on the organizational and IT-related constraints (see Figure 10.4).[282]

10.2.1
To-be modeling based on an ERP-reference model

The first alternative is a procedure that focuses on the related ERP product. It allows a quick modeling by adopting a basic ERP solution which then "only" needs to be configured, and which ensures a timely use of identical ERP terms. There is a further guarantee that these models can be supported by the software. In addition to the costs for software-specific reference models that might have to be purchased separately, the following problems may arise:

Problems with reference models

- In most cases, it is not possible to compare the ERP-specific models with the standard software models of other vendors.
- The current as-is situation and the ideal (planned) situation of the organization are only taken into account based on intuition and experience.

---

[282] See also Rosemann, Rotthowe (1995), p. 11.

Analogous to the discussion about the possible counter-benefit of as-is modeling effort (see Chapter 5), there is a potential risk that too much emphasis is placed on this reference model when orienting toward a specific ERP-solution, and the individual requirements of the company are neglected.[283]

Process alterations

The downfall of this procedure is the tendency to force process alterations, since the adopted process is based on predefined process models from the provider. The less restrictive the software, however, the less this situation occurs, as the range of possible parameters increases. This provides for a greater number of scenarios to adopt, reducing the need to make alterations to current processes.

10.2.2
Modeling the perfect case

This strategy dramatically emphasizes the individual organizational requirements. The outcome of this modeling activity is an "ideal model" that is widely independent of software restrictions but which actually includes organizational restrictions that assume to be subject to software alterations. This ideal model is taken as a basis for selecting the software. In order to judge the suitability of a certain ERP-software for company-specific objectives, it must be possible to compare the models. A procedure must, therefore, be established for identifying and comparing aspects such as the components of the information model; the model's width and depth (degree of detailing); the scope of presented special cases; and the attributes. A technical term model, for example, can take into account problems of missing general naming conventions by explicit maintenance of homonyms and synonyms that result from the difference between company terminology and ERP terminology. The to-be model results from comparing a company-specific ideal model with the ERP reference model. To the same extent in which the independently-created model must be edited in order to be comparable with ERP-software models, the freedom to select the modeling method, the degree of detailing, or the modeling scope are restricted as well.

Ideal model

Software

---

[283] See Reitzenstein (1998), p. 393.

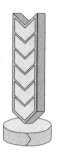

10.2.3
Generation of to-be models based solely on current as-is models

Waiving ideal models

The third alternative is to waive ideal modeling since it characterizes a state that can only be reached at tremendous cost. The to-be model is compared with the ERP-reference model based on as-is modeling. The more the intensity of as-is modeling is reduced, the more this alternative approximates alternative 1.

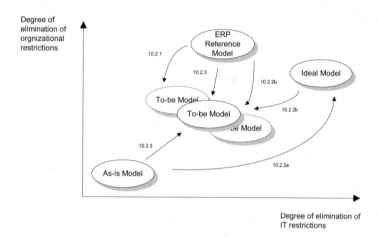

Fig. 10.4. Alternative approaches when using ERP-reference models

10.2.4
Design recommendations

The following design recommendations can be given for modeling projects where process management targets are considered and the implementation of ERP-systems is planned:[284]

Design Recommendations

- The business domains, where ERP-software must be used or is already in use and is also documented by reference models, must be explained at an early stage. This explanation identifies the domains where interdependencies between organization-centered and software-documented models are expected.
- The method experts involved have to be trained in using the functions of the reference model that are offered by the soft-

---

[284] See Rosemann, Rotthowe, Schütte (1999), pp. 212-213.

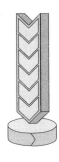

ware. The focus should be on method and tool aspects rather than on the business contents. This will guarantee that the method experts take responsibility for all the contents of the relevant models themselves.

- The employees who are in charge of the implementation and handling of the software must be trained as well, because the application of reference models for the time being is more the exception than the rule. Although reference models are an integral part of the SAP R/3 software, for example, they need not be bought separately.

- A modeling project should be based on conventions (see Chapter 3) that take into account the modeling conventions of the software. The actual impact of the ERP-reference model and its conventions depends of course on the comprehensiveness of the ERP-use.

In the case of DeTe Immobilien, the company does not regard the selection and implementation of ERP-software as the main purpose of the modeling project. Nevertheless, reference models of ERP solutions used in the company were implemented in the modeling project.

The reference models in the ERP-software were taken as reference points for the identification of potential processes, in particular in the administrative areas of financial accounting, cost management, human resource management, and parts of material management.

Beside the pure process identification, the process models themselves became part of the work of the related process teams. Here, two procedures had to be distinguished:

- As far as as-is and / or to-be models are mainly determined by ERP-software, related reference models were immediately taken as basic solutions for modeling activities. An example is the cash management process whose basic features were identified by manually-entered markings in a printout of the reference model from the SAP R/3-Software. The final, modeled cash management process mainly represented a subset of the process that could be potentially covered by the software. This procedure can be recommended above all when the participating expert users have a detailed understanding of how to use the software and how to execute the corresponding organizational tasks. When individualizing these models, the participating organizational units, the (ERP-external) application systems, as well as the technical terms for input and output, have to be added.

Use of R/3-reference models

Integrity test using ERP-reference models

- Beside this method, which directly orientates itself toward a software-specific reference model, it is also a proven practice to retain the ERP-reference models for the present. In these cases, the process models were generated without any reference to the basic ERP-solution. The moderator, however, should know the contents of the relevant ERP-reference models and should take them as a guideline. Immediately after termination of the first solution, the models were compared with the ERP-reference models. This comparison served as an integrity test and as a catalyst to discuss alternative solutions. In addition, this comparison was used in the for to-be modeling phase to learn which software adaptations were required, and if they were too expensive, as well as which functions had to be taken to adapt them to the ERP-software. This procedure was followed by the generation of processes for accounts receivable and accounts payable in the payment reminder system. In this case, the analogous structure of both corresponding ERP-reference models was also an impetus for a uniform design of both reminder processes in the modeled company. Consequently, these processes were assigned to a process owner.

In addition, selected users of ERP-software, but also project members who were responsible for the implementation of ERP modules, as well as members of the modeling teams, were all informed about the options offered by software-specific reference models (which were often unknown, or not used sufficiently) and trained accordingly.

10.3
Development of in-house software

10.3.1
In-house and off-the-shelf software

In-house software and off-the-shelf software

Software systems can either be bought or implemented as off-the-shelf software or as self-developed individual "in-house" software. In the 50's, 60's and 70's, widely-in-house software was predominantly in use, while in the following years off-the-shelf software has been given preference. This is explained by the tremendous cost resulting from the development of in-house software, and, above all, from the maintenance cost of self-developed software. The lack of engineering involvement for software development, as well as the insufficient handling of the business economical prob-

problems by the programmers, led to the so-called software crisis that was characterized by an untimely production of programs with insufficient functionality. The high maintenance cost could be explained by a poor mapping of user problems and by a lack of flexibility in the implementation. Added to this was the fact that the beginning of the software development process is often intuitive. This procedure did not meet the requirements of a more industrial production of software in terms of exact phases and predefined results.

Software crisis

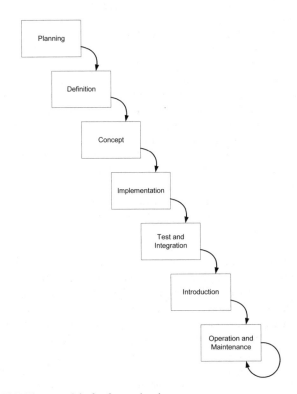

Fig. 10.5. Phase model of software development

In the meantime, software engineering has developed and has become a mature research domain which offers quite a number of procedure models for software development which support the total life cycle of software systems, e.g., the "waterfall model" of ROYCE[285] and the "spiral model" of BOEHM[286]. The different proce-

Procedure models

---

[285] See Royce (1970).
[286] See Boehm (1988).

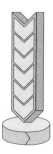

dure models are distinguished by the number and types of individual phases in the development process and, in part, in their orientation toward different programming prototypes. In principle, the following phases can be found in nearly all procedural models (see Figure 10.5).[287]

10.3.2
Design of in-house software with process models

Through the stronger formalization of software development and due to the increasing requirements of software systems with an increasing complexity, caused by the individual needs of companies, it is recommended to use process models for software development in order to describe the requirements of the product to be created with sufficient precision and integrity.

Planning phase

In the planning phase, a feasibility study of the development project is executed. Alternative procedures for the following phases are investigated. In the planning phase, business process models can be used to identify functions and / or activities to be supported by software. Process models simplify the creation of specifications, since the participating organizational units and the data streams to be processed can be taken directly from the process model if the model is sufficiently detailed. Here, the process models serve as a basis for defining the professional requirements of the software system.

Definition

In the definition phase, the data to be processed is defined as well as the functions to be executed by the software system, the dynamics, (i.e., the logical process), and the user interface. Based on the data streams in the process models, elementary data can be identified for individual applications. A summary of sequential functions executed by the same organizational unit enables the adaptation of the logical flow of the in-house software to the requirements of the business processes in the company.

Concept phase

In the concept phase, the data and functions from the former step are specified on a formal conceptual level (in the form of entity relationship diagrams, activity diagrams, function trees, class diagrams, etc.). The specification can be checked for correctness using the existing data in the process models.

Implement-ation phase

The implementation phase serves to code the specification from the concept phase for practical use. The individual elements of the generated software (modules, objects, tables, etc.) are created in a related programming environment and the ready-made components are subject to single tests (white-box, black-box, etc.).

---

[287] See Balzert (1996).

In the test and integration phase, the individual modules from the implementation phase are integrated into the total system and checked for interface compatibility and errors via integration tests.

Test and integration phase

The purpose of the introduction phase is to transfer the ready-to-use software system from the development environment to the production environment, to train the users in the handling of the new system, and the transferal of operative databases to the new software system.

Introduction phase

In the operation and maintenance phase, the software system is put into operation and continuously adapted to the changing environmental conditions (e.g., increased tax rate) and / or will be updated with new functions to be provided (e.g. support of Internet orders) in the course of completion and adaptation maintenance.

Operation and maintenance phase

It can be stated that it is sensible to use process models to create in-house software, particularly in the analysis and definition phases, and to specify the requirements of the software system to be created with the necessary precision, the required verification, and without redundancies.

10.4
Workflow management

10.4.1
Motivation for workflow management systems

In addition to the conversion of an organization focused on functional silos into a process-oriented structure, it is also required to provide adequate IT-support for the redesigned business processes. The major part of today's application systems, however, cannot cope with the requirements of process orientation since they are constructed in the form of functional programming hierarchies and, therefore, orient themselves toward the fulfillment of single functions. Workflow management offers the possibility of integrating these functionally-structured applications in process-oriented applications by guiding the users through the system when executing business processes and by providing them with the necessary data and applications.

IT-support of potential optimizations

While the development of information system architectures separated databases from application logic in the 70's, workflow management separated the management of the control flow from the application logic in the mid 80's (see Figure 10.6). Here, application-dependent services such as the persistent exchange of messages between application systems or application control via proc-

From data integration....

...to control integration

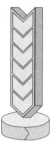

ess models, should be integrated into a system-wide layer, the so-called middleware.

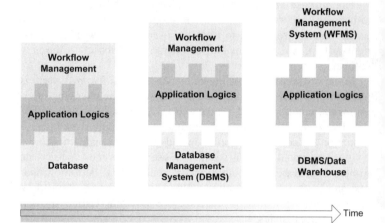

$\longrightarrow$ Time

Fig. 10.6. Development of information system architectures in the course of time (according to Rosemann, zur Mühlen (1998))

10.4.2
Related Definitions

Workflow

A workflow is that part of the process that contains the timely and logical sequence of activities of a job as well as information about the data and resources that are involved in the execution of this job.[288] The objective of a workflow is the automated process execution where the transitions between the individual activities are controlled by the workflow management system (WfMS). Data that influences the sequence of activities is designated as workflow-relevant data (e.g., the amount of a credit that is determined by the administrator, in contrast to the activity-internal data that is designated as useable data (also: application data) (e.g. letter about acceptance or rejection of a credit request, see Figure 10.7).[289] Internal control data of a workflow management system is designated as workflow-internal data. Workflow-internal data includes information about the initiator of the workflow and / or start and end dates of single activities. Consequently, workflow management means coordination and control of the workflow.

Useable data, workflow-internal and workflow-relevant data

---

[288] For the following, see Becker, zur Mühlen (1999).
[289] See WfMC (1999).

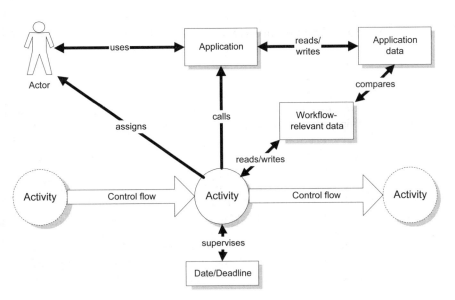
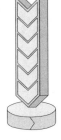

Fig. 10.7. Workflow activity

A workflow management system is an application system that en- *Workflow*
ables and / or supports workflow management. Some workflow *management*
management systems include additional tools to generate work- *system*
flow models while others access existing tools for business process
modeling via interfaces (e.g. ARIS).[290]

A "workflow application" is a combination of a workflow man- *Workflow*
agement system and embedded application systems that are used to *application*
support a process that is based on a workflow model. *and invoked*
application

An "invoked application" is an existing software solution that is
embedded in one or more activities of the workflow model from
which it is called up at runtime in order to enable the user to do his
job. This can be a word processing system, but also a mainframe
application or a transaction in an ERP-system. The data is exchanged between the workflow management system and the invoked application, which is of importance for the further course of
the workflow instance. The useable data that is processed by the
individual application programs is transparent to most of the work- *Process-*
flow management systems in the sense that it is "invisible". An *oriented*
exception is the so-called process-folder system, where a folder is *systems vs.*
electronically mapped and routed through a network of worksta- *process-folder*
tions. Systems that follow this principle are, for example, *oriented sys-*
CSE/WorkFlow or IABG ProMiNaND. *tems*

---

[290] See Rosemann, zur Mühlen (1998), zur Mühlen (1999).

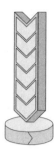

Specification of resources

Role = qualification

Some workflow management systems allow the workflow modeler to specify additional supporting applications (helper applications) that can be called up by the user, if required, but which, however, are not absolutely necessary for the processing of the task. [291]

Workflow activities specify the resources that can take over the execution of the related activities at runtime. Resources can be employees in the user departments or software resources and / or machine resources. Normally, this occurs within the framework of role specification. Here, an abstraction is made from specific employees so that every activity is assigned the required qualification to execute this activity. This qualification is called a "role". In parallel with the workflow model that is specified in this way, the individual resources are assigned to the related role in a (separate, if applicable) organization model. At runtime, the workflow management system determines the resources that are qualified to execute the activity based on the specified role and organization model, and assigns the corresponding work step to these resources.

10.4.3
Transfer of business process models into workflow models

Workflow projects

Adaptation of functions

Specification of data

Normally, workflow management projects are started after business process modeling and / or reorganization projects. The developed business process models are used as source documents when creating workflow models. The objective of workflow modeling, however, is different: While business process modeling focuses on the organizational design, workflow modeling concentrates on the IT-support, i.e., the available process models have to be adapted to the workflow management purposes.

First, the functions must be adapted to the granularity of an IT-application. In this connection, the functions that are supported by an application can be summarized in a functional block. If a function needs to be supported by multiple application programs, it must be further refined, if applicable, i.e., to be replaced by multiple functions of finer granularity.[292]

The required data must be specified for every function. Above all, data which determines the process flow (workflow-relevant data) has to be identified. Depending on the workflow management system support, such detailed data must be collected which is then transferred between the activities of the workflow management system. In the simplest case, a reference is made to centrally-

---

[291] See e.g. IBM (1998).
[292] For a discussion of granularity of applications in workflow management systems, see Becker, zur Mühlen (1999).

stored data, e.g., an invoice in a document management system. In complex cases (e.g. when the workflow management system has to secure the integrity of useable data), all data that are needed to execute a function must be defined. This requires defining and implementing the related data types and entity types.

The conditions for branching out (alternatives, side runs) in a process model must be described in such a way that they can be evaluated automatically. While in a business process model information such as "credit to be checked by board of directors" may be sufficient, this condition must be translated in the workflow model into detailed criteria such as "credit amount is greater than US$ 30,000" and "credit amount is smaller than or equal to US$ 30,000".

Description of workflow conditions

For every activity it must be determined who is authorized to execute this activity. In most workflow management systems this is done via "roles", i.e. concepts which separate the actual person from the activities of the workflow model. Roles are defined for every activity. The role then has to be assigned to the executing person (e.g. "member of buying department", "authorized to sign contracts above US$ 3,000", "English language"). At runtime, this information is compared with the actually-available executives, and those people who fulfill the qualification conditions will be informed of the pending tasks (role solution).

Determination of executives

Activities in workflow models can be supported, in full or in part, by applications. In this case, it must be defined which program has to be called up to execute this activity. This information includes the path to the program, the parameters to be transferred (derived from data definition) and the expected parameters for results.

The so-designed workflow model can be implemented in the basic process model via feedback engineering or business activity monitoring. In particular, when weaknesses become obvious or when ambiguities need to be eliminated, a revision of the basic model is sensible. Figure 10.8 summarizes the transition steps as described above.

Feedback-Engineering

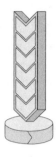

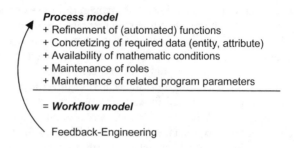

Process model
+ Refinement of (automated) functions
+ Concretizing of required data (entity, attribute)
+ Availability of mathematic conditions
+ Maintenance of roles
+ Maintenance of related program parameters

= **Workflow model**

Feedback-Engineering

Fig. 10.8. From process model to workflow model (source: Galler (1997))

10.4.4
Identification of business processes suitable for workflow management

*Busines
process
≠ workflow*

Not every business process is suitable for support by a workflow management system. In order to select from the potential processes for workflow automation the ones that show high success probability for implementation and the best economic promise, a multi-dimensional criteria catalog should be implemented.[293] With the help of this criteria catalog, processes can be evaluated regarding their for workflow suitability utilizing three evaluation categories.

*Criteria to
identify work-
flow- suitable
processes*

- Structured features of the process (e.g., evaluation of process-oriented, cooperating resources [participating user department, application systems, databases], structure of process, organizational maturity, continuity, process quality, etc.),
- Organizational conditions within the framework (e.g., innovative mentality of involved user departments, timely availability of expert users, strategic importance of process, obligation for co-determination and existing documentation),
- Potential benefits of the workflow-automation for the process (contribution to corporate goals, definite need for action, reduced run-time, increased process transparency, digitizing of routine tasks, and modularization of application systems).

*Adaptation of
evaluation
scheme*

Figure 10.9 shows an extract of such catalog criteria. This evaluation scheme can be adapted to the specific needs of the company on three levels. Every elementary criterion (e.g., number of participating user departments) is provided with a weighting factor. The elementary criteria are summarized into a group that in turn is

---

[293] See Becker, v. Uthmann, zur Mühlen, Rosemann (1999).

weighted (e.g., documentation of process). Finally, the groups are collected in individual evaluation categories that in turn can be weighted separately. This way, the present evaluation scheme can be adapted individually.

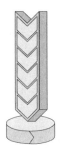

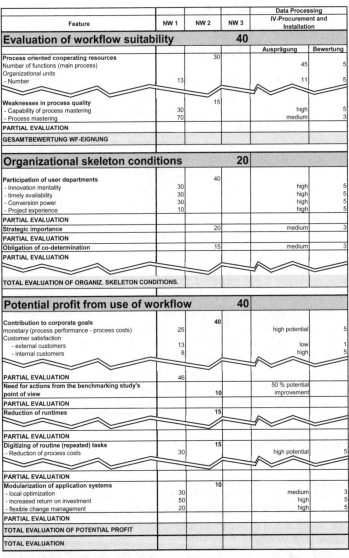

Feature	NW 1	NW 2	NW 3	Data Processing IV-Procurement and Installation	
Evaluation of workflow suitability			**40**		
				Ausprägung	Bewertung
Process oriented cooperating resources		30			
Number of functions (main process)				45	5
Organizational units					
- Number	13			11	5
Weaknesses in process quality		15			
- Capability of process mastering	30			high	5
- Process mastering	70			medium	3
PARTIAL EVALUATION					
GESAMTBEWERTUNG WF-EIGNUNG					
Organizational skeleton conditions			**20**		
Participation of user departments		40			
- Innovation mentality	30			high	5
- timely availability	30			high	5
- Conversion power	30			high	5
- Project experience	10			high	5
PARTIAL EVALUATION					
Strategic importance		20		medium	3
PARTIAL EVALUATION					
Obligation of co-determination		15		medium	3
PARTIAL EVALUATION					
TOTAL EVALUATION OF ORGANIZ. SKELETON CONDITIONS.					
Potential profit from use of workflow			**40**		
Contribution to corporate goals		40			
monetary (process performance - process costs)	25			high potential	5
Customer satisfaction					
- external customers	13			low	1
- internal customers	8			high	5
PARTIAL EVALUATION	46				
Need for actions from the benchmarking study's point of view		10		50 % potential improvement	
PARTIAL EVALUATION					
Reduction of runtimes		15			
PARTIAL EVALUATION					
Digitizing of routine (repeated) tasks		15			
- Reduction of process costs	30			high potential	5
PARTIAL EVALUATION					
Modularization of application systems		10			
- local optimization	30			medium	3
- increased return on investment	50			high	5
- flexible change management	20			high	5
PARTIAL EVALUATION					
TOTAL EVALUATION OF POTENTIAL PROFIT					
TOTAL EVALUATION					

1=low to 5 = high suitability for WF

Fig. 10.9. Catalog of criteria to evaluate workflow suitability (extract)

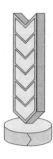

Mandatory criteria

Before processes pass such detailed analysis it is recommended to reduce the number of processes to be analyzed to approximately five to seven by analyzing a few key or mandatory criteria. Such mandatory criteria can refer to, among others, the strategic importance of a process, the availability of user departments involved, but also the maturity of the process.

10.4.5
Execution of workflow models

Information about work lists

Audit Trail

When executing specified workflow models, qualified employees are informed about the status of pending tasks via so-called "work lists". If more than one employee is authorized to do this work, a related entry is made in the work list of all qualified employees (this entry is called "work item"). As soon as one of these employees selects the job, the entry (work item) is removed from the work list of the remaining employees. The workflow management system calls up the application system which belongs to this work (e.g. a SAP-transaction for "invoice entering"), provides the required data, and logs the start and end date of this work in a log file that is called an "audit trail". After termination of the application system that executes the master control, the process control returns to the workflow management system. The workflow system evaluates the returned data and continues with the execution of the next activity and / or with the evaluation of the next flow control condition (branching or consolidation).

If an exceptional situation occurs in the execution of a workflow (e.g., if no employee selects the work item within a predefined timeframe or if an application system does not deliver valid return codes), an escalation procedure comes into force. In this case, the process- or activity responsible is informed, who must then deal with the workflow execution that is either resumed or aborted.

10.4.6
Potential benefits from workflow management systems

Coordination

Workflow management systems primarily take over coordination tasks in the electronic support of process execution. In addition to the coordination of activities, the application systems involved, the authorized employees, and the relevant data are coordinated as well. From these coordination functions the potential benefits of workflow management systems can be derived:[294]

---

[294] See Rosemann (1998c).

Through the coordination of activities, the manual activities to transfer data, documents, and notes between the workstations can be omitted. Related transportation times are, thus, eliminated. Furthermore, the automated routing of documents reduces the variance in the execution of a process, i.e., in a workflow-supported process, and equal process objects are always processed the same way. This contributes to process mastering and process quality improvement.

Elimination of manual coordination functions

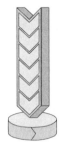

The coordination of application systems enables the workflow management systems to execute a modular design of the IT landscape of an enterprise. Since monolithic application blocks are no longer required, problem-specific applications of finer granulation can be used. These are often more efficient and less expensive and, therefore, more cost-effective and easier to maintain than complex, packaged solutions. Some workflow management systems offer additional functionality in the form of script languages and form designers. If these tools are powerful enough, it is most likely possible to manage the workflow without using external applications, i.e., the supported process can be solely controlled by the workflow management system (see Figure 10.10).

Workflow-based CASE

Formular : Antrag Prüfen

Formular Ende Hilfe

Vorgang: ORDER
Case: 11-1 Order Mizu

Last name: zur Mühlen   Date of application: 30/11/1998
First name: Michael
Job title: Research Assistant
Department: OE134
Room: 032

Equipment type: Laptop   Total Value: 5,000.00
Network node: 312410
IP Address: 128.176.158.85

Authorized   YES
                 YES
                 NO
Memo:

Fig. 10.10. Workflow management system with mask editor

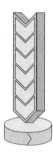

Balanced worksharing

The automatic assignment of activities to task carriers in connection with the resolution of roles, leads to a balanced distribution of work among employees with the same qualification, since an activity is offered to all related employees for processing. By eliminating routine tasks, such as the handling of internal mail, the employees can concentrate on higher-ranked activities. This can result in an increased employee satisfaction.

Data integration

The coordination of relevant data for process execution by a workflow management system supports the identification of data redundancies, the consolidation of process-related data in a central database, and offers the possibility for further workflow mining and monitoring.

10.5
Knowledge management

10.5.1
From process management to knowledge management

Knowledge as production factor

The topic "process orientation", which dominated the discussions in the early 90's, has been replaced in recent years by the topic "knowledge management". The basic aspect in these discussions is the recognition that the industrialized nations develop from a production-orientated industrial society where the major production factors are "material and money", into an information and knowledge society where the most important production factor is "knowledge". Knowledge management deals with the question of how to use the available know-how and how to find organizational structures that are suited to the support, creation, distribution, and maintenance of this knowledge. In most cases, there is no clear differentiation between the terms data, information, and knowledge.[295]

10.5.2
Data, information and knowledge

Data, information and knowledge

Data is mainly defined as the collection of single, isolated facts that need to be interpreted by the user.[296] Information is defined as interpreted data that is helpful for the user to solve a specific problem. Knowledge, within this classification, is defined as the capability to interpret data and to convert the interpreted data into information.

---

[295] For a discussion of these terms, see e.g. Schütte (1999).
[296] See Tuomi (1999).

Knowledge in a company exists in two different forms. On the one hand, numerous, explicit sources of knowledge exist in a company, such as handling instructions, quality assurance prescriptions, activated correspondence, etc. On the other hand, the employees of a company have at their disposal an implicit knowledge, such as business practice, corporate culture, or certain rules of conduct. This is also known as tacit knowledge.[297] While the explicit knowledge is available in coded format and can, therefore, be easily distributed and made known to others, tacit knowledge requires its contents to be formulated in order to distribute this knowledge within the company. Consequently, the objective of knowledge management is to convert both forms of knowledge into a format that can be used by the company.

Explicit and implicit knowledge

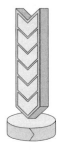

10.5.3
Process of knowledge creation

The absorption of explicit knowledge by employees, the explication of tacit knowledge in writing or another form of representation, as well as the related transformation of both forms of knowledge, cover the four fields of knowledge procurement according to NONAKA and TAKEUCHI. (See Figure 10.11). Every presented piece of knowledge can be transformed into the same or into another field of knowledge so that a closed loop results.

4-phase model of knowledge creation

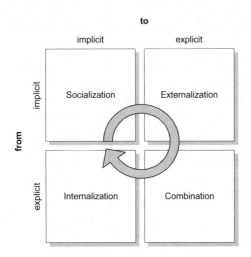

Fig. 10.11. Process of knowledge creation (source: Nonaka, Takeuchi (1997), p. 75)

---

[297] See, for example, B. Holsapple, Joshi (1999), p. 3.

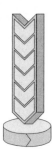

Socialization

Socialization is the transformation of implicit knowledge of a knowledge carrier into the implicit knowledge of another knowledge carrier, e.g., the introduction of a new employee into a workgroup where he adapts to the "unwritten laws" of this group and transforms this knowledge into his own principles, or where experiences are exchanged between colleagues with different experiences in the execution of an activity.

Externalization

Externalization is the specification and publishing of implicit knowledge. This, for example, is done when a common business practice is formally specified for transformation into an IT system. The documentation of processes in a company falls under this topic also, if, formerly, no other documentation was available, for example, in the form of procedural instructions. In addition, new knowledge can result from an analogy of separate areas, when, for example, production processes from another branch are transferred to the original branch. NONAKA and TAKEUCHI give the related example, in which the production of a canned drink is transformed into the production of aluminum platens for photocopiers.[298]

Combination

Combination is a collection of different types of knowledge, for example, the development of a new product in which employees from different departments participate. Here, documents, files, and other information carriers are combined, classified, and stored in a new combination.

Internalization

Internalization of explicit knowledge means that employees practice new process models until they no longer need the formal description of the processes as reference and feel the necessary working steps to be logical and, therefore, acceptable ("learning by doing").

Knowledge-based process life cycle

The process of knowledge procurement is repeated and leads to a spiral where original, individual knowledge is transferred to other individuals until the level of a group is reached and, finally, until the level of the total organization is reached. With reference to process modeling, the process of knowledge procurement can be described as similar to the life cycle of a process model, including the four phases: generation, modeling, consolidation, and conversion. This form of a life cycle is shown in Figure 10.12.

---

[298] See Nonaka, Takeuchi (1997), p. 78.

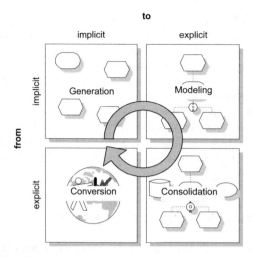

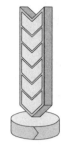

Fig. 10.12. Knowledge-based process management

10.5.4
Organizational support

The efficient procurement and use of knowledge in a company requires that these activities are supported by appropriate organizational structures. This includes that employees are encouraged to
learn and to pass their experiences onto others. Those supporting
activities include the establishment of forums in the form of ple-
nary sessions and the organizational anchoring of knowledge
transfer, e.g., in the form of documentation guidelines. The documentation guidelines should not only determine the form of documentation but, above all, the process of continuous documentation
of project results. The exchange of knowledge in a company can
also be supported by advanced training programs, job rotation, and
by incentives that promote the cooperation of user departments.

Establishment of exchange forums and documenta- tion guide- lines

10.5.5
Information-technological support

In addition to organizational measures to promote knowledge
transfer, information and communication technologies are essential
tools for the collection and distribution of knowledge in a company. In particular, in externalization and combination phases,
modern information systems contribute to an acceleration of
knowledge transfer and to an increase in information quality.

Intranet and groupware

Document management systems

Examples of such tools are the Intranet where the employees can file their reports and documents, and where they can access information from other employees, departments, and user domains. GroupWare and similar systems promote the informal exchange of information through discussion forums and electronic "black boards". Archive servers and document management systems offer transparent access to historical data and can be used in the combination phase. Data-Warehouse technology, OLAP, and data mining technologies also contribute to the identification and explanation of previously unknown relations between the data of an enterprise.

10.5.6
Implications for company management

Knowledge management is a management task

Knowledge management extends the scope of responsibility of the management. The knowledge of individuals is no longer a demand at the employee level but must become a task at the management level. The requirements of this task are to create guiding principles, to anchor the knowledge transfer in the organizational structure of the company, and to provide the company with the required incentives and technical infrastructure for an efficient and effective know-how transfer.

10.6
Model-based benchmarking

Definition of benchmarking

Process models generated in reorganization projects can be further used to evaluate the performance of a company and to eventually initiate changes in its business practice. The comparison of actual business data and practices with guidelines and practices is known as "benchmarking".[299] The objective of benchmarking is the comparison of companies, parts of a company, or processes in order to identify, analyze, and describe a perceived best practice.[300] This practice is then adapted to one's own company.[301] In this connection, two types of benchmarking are distinguished: Benchmarking within a company, and benchmarking with external organizations.

---

[299] See e.g. Meffert (1994), p. 152.
[300] Also see Chapter 5.3.3.
[301] See Lamla (1995), p. 28.

10.6.1
Procedure model for benchmarking projects

CAMP describes the benchmarking-process of the XEROX Company, which is taken here as an example for a benchmarking project.[302] In the planning phase, the relevant benchmarking object is
determined as well as its results and the recipients of these results.
For example, there can be a determination to analyze the order
processing process in regard to the processing time, where the output is defined as a delivery of ready-made products to the customer. For this investigation object (i.e. the order processing process), comparable companies are searched which either belong to
the same field as the company investigating the benchmarking, or
may also come from other areas. In the latter case, however, the
compared object in the compared company must show similar features as the object of the investigating company. Therefore, the
number of comparable companies for generally administrative
and / or supporting services is larger than the number of comparable companies for concrete product-related parameters. In addition,
the methods for comparison and the procurement of relevant data
are defined in the planning phase.

Planning phase

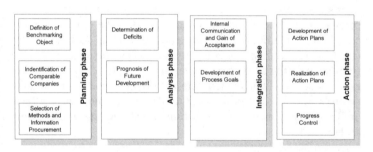

Fig. 10.13. Phases of benchmarking process (in adaptation to Camp (1989) as
well as Lamla (1995), pp. 24-26)

In the analysis phase, the qualitative and quantitative deficits of the
investigating company are determined by a deviation from what is
considered "best practice". In other words, how differently the
company performs the order processing process, for example,
compared to other companies investigated. The reasons for qualitative deficits are to be found in different processes that are executed
to achieve the same result. Quantitative deficits result if the comparable companies create different parameters within a process,

Analysis phase

---

[302] See Camp (1989).

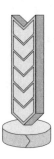

Integration phase and action phase

such as transport, idle or processing times. Since the deficits are time-related, the next step contains a projection as to how the comparable partners will develop with reference to the compared object, for example, by market changes or new production methods, in order to consider this development in the integration and action phase.

In the integration phase, the results of benchmarking research are distributed within the investigating company in order to gain acceptance by the related user departments, as well as to determine the subject-related process goals to be executed in the action phase.

10.6.2
Benchmarking within a company

Comparison of different parts of the

Company internal best practice

Benchmarking within a company allows the company to compare the actual business practice with the instructions documented in the process models. When a part of the company (for example, a department) is compared with the documented process models, benchmarking approximates the classical controlling. It is, however, also conceivable to compare foreign company units with each other when they perform similar processes in order to identify internal best practice. This, however, requires related attributes and parameters to be collected and documented in the individual company units. If process modeling projects contain as-is processes (see Chapter 5), then reorganization effects can be directly derived from the existing as-is data – as far as such data is available. If no as-is models and / or related data are / is available, then periodic controls can be executed in order to identify learning curve effects in the same processes, such as continuous shortening of processing times.

10.6.3
Benchmarking beyond organizational limits

"Creative" benchmarking

Within the framework of the organization-wide benchmarking the objectives of a company can be compared with the objectives of its best competitor. But comparisons with companies and / or processes in other industries can also provide useful information for the further development of one's own company. Thus values from other industries can be utilized for the identification of optimization potentials within the framework of "creative" benchmarking. An example for this is given already in Chapter 10.5, where the production of canned drinks is compared with the production of aluminum platen for copiers and laser printers. If the functional ar-

eas of companies are compared with each other, the number of potential comparison partners is greater than when comparing function-wide objects such as complete business processes. In this
case, the comparison partners have to be analogous in structure, a
fact that makes a comparison with companies from different industries difficult.

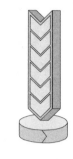

A critical factor in the comparison of processes beyond organizational limits is the availability and quality of data. Since, normally, internal information, such as idle times and transport times
of process objects, is not revealed to third parties because of their
strategic importance, it makes more sense to compare processes in
direct contact with the customer by collecting the data of the compared company from the functions: inquiry processing, order con-
firmation, complaint management, as well as invoicing and payment modalities. Since the customer can discern this data
immediately, this data is of great importance for one's own business processes. A different evaluation applies for processes seen
start of first activity to end of last activity), idle time and / or wait
time (time that a process object stays in a process without being
worked on), transport time (transfer time from one individual to
another), and processing time (time of active work on a process
object). In addition to these absolute values, the parameter variants
are also of interest since they can be indicators for insufficient
process mastering. If, for example, the processing time shows a
high variance, it should be analyzed whether or not modifications
in the process flow can reduce these variants.

Problem:
Availability
of data

Analysis of
customer data

10.6.4 Comparable parameters

Benchmarking divides the parameters into two dimensions – quality and quantity, as well as into three categories: time parameters,
quantity parameters, monetary parameters, as well as other parameters. These parameters can be collected from a functional a-
rea, from a process, from an activity, from a special product, or
from a participant (executive) and / or resource.

Quality and
quantity
dimensions

Time parameters to be used for a benchmarking process are:
process runtime (time between start of first activity to end of last
activity), stay time and / or wait time (time that a process object
stays in a process without being worked on), transport time (time
from one individual to another), and processing time (time of active work on a process object). In addition to these absolute values,
the parameter variants are also of interest since they can be indicators for insufficient process mastering. If, for example, the processing time shows a high variance, it should be analyzed whether
or not modifications in the process flow can reduce these variants.

Time
parameters

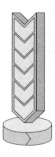

Quantity parameters

The quantity values that can be of importance for process bench-marking are input-output relations, i.e., the relation between the achieved results and the used resources. These values may refer to the number of employees involved in the process, material resources, or information about the quality of process objects. Information about the error rate of a process (the relative share of erroneous process instances), the error types (number of different errors occurring during process execution), the error rate (absolute share of erroneous process instances) and error frequency (timely distribution of two sequential errors) are indicators of poor process mastering. Changes to these values in the course of time are indicators of either learning effects or decreasing attention of the process participants. Error situations include not only defective goods in production, but also include delays of delivery, share of unexpected questions, share of stopped processes (for example, because of order cancellations) and external effects such as delayed deliveries from sub-suppliers.

Analysis of error types, error rates, and error frequencies

Monetary parameters

Monetary parameters in process benchmarking can be the output from generated process objects in the form of achieved turnover, and the input in a process object in the form of costs, such as material and personnel costs.

Other parameters

Other values of interest for process benchmarking are the number of process activities, the number of site changes, the position of interfaces to other processes (e. g., early or late archiving of incoming invoices in the audit trail process) and, if applicable, media breaks in a process, i.e., the number of changes between different media on which the process object is stored, i.e., on paper, disk, or CD-ROM.

The indicated values can be compared both on the process level and on the activity level as far as the related data can be collected with a reasonable amount of time and cost. A comparison of similar functions in different processes may point to an organizational environment which is best suited to execute a function, i.e., it indicates whether a project-oriented organization is more suitable for a certain type of task than a function-oriented organization.

10.6.5
Measurement and strategic influence of activities for reorganization

Collection of parameters

When parameters are to be compared for benchmarking, this can only be done if these parameters are collected in a reliable manner. The higher the strategic importance of activities, the more difficult the measurement of these activities and / or values will be. While time and monetary effects, such as a reduction in personnel costs or shortening of runtimes, can be measured and compared directly, the answer to questions of whether or not a faster market entry for a developed product is advantageous for the company is by far more difficult to quantify. The expansion of the process participants' scope of tasks (job-enrichment, job-enlargement) can be expressed by measurable parameters (e.g., personal efficiency in handling process objects within a defined period of time / time unit). The evaluation of these effects, however, cannot be done without considering the framework conditions. Example: A decrease in routine activities is regarded as positive when the employees are actually able to execute higher-ranked tasks that contribute to the success of the company. If, however, no higher-ranked tasks exist, then the release of potential HR resources cannot be regarded as a measurable profit for the company. Figure 10.14 shows this contradicting effect in graphical format.

Measuring of effects does not require their strategic importance

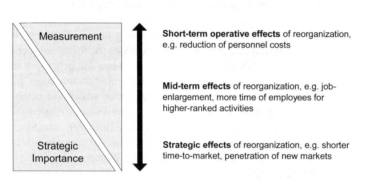

Fig. 10.14. Measurement vs. strategic importance of reorganization

10.7
Internet-based process modeling

10.7.1
Motivation of decentralized process modeling

The use of modern communication technologies such as mobile phones, notebooks with modems, and local Internet access goes hand in hand with an increased decentralization of company activities. Representatives in the field who are provided with a laptop and a modem do not need a central office for the electronic transfer of contracts, inquiries, orders for information material, and / or

Decentralized company activities

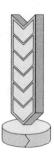

Teleworking and tele-commuting

calendar data. The existing office has a changed function and now serves as a communication center, location for an exchange of opinions with colleagues and as a social meeting point. Some enterprises abstain from assigning a firm workplace to their employees who are working in the field; instead the available workplaces are assigned in the morning according to the "first come – first serve" principle. The personal effects of the employee are stored in moveable containers or storage boxes. In the literature, the tendency for decentralization of previous workplaces is treated under the catchword "teleworking" and / or "tele-commuting" for "workflow automation".[303]

The ongoing development of communication technologies in the sense of larger bandwidths (i.e., a higher quantity of data transferred by time unit) favors the generation of virtual companies, where the individual organizational elements implement their core competencies in the virtual company for an (often) limited period of time (e.g., for the period of a defined project).

Modeling by project teams

The processes are distributed for processing per se. The knowledge of the individual activities in the processes is, therefore, centrally stored in the executing business units. When the processes in a company are to be modeled, this central information must be consolidated. This can be done by project teams which are established and who meet at a central location to hold workshops or interviews.[304] These central modeling sessions, however, are often time and cost consuming. Therefore, the search for alternatives is intensified in order to reduce the time of the expert users and the cost for business trips.

Modeling via Internet

Three scenarios

A suitable option is the use of Internet technology that allows a decentralized modeling of business processes based on a central process database. The widely distributed TCP/IP protocol (Transmission Control Protocol / Internet Protocol) represents the common communication base of computers in the Internet, and enables distributed workgroups to cooperate via Internet using a large number of applications. This procedure will be explained as follows by taking three scenarios as examples. In the first scenario, the process models are generated on a central server that the expert users access on decentralized modeling clients via the Internet. In the second scenario, the process models, which are generated on a decentralized basis, are sent to a centralized moderator via the Internet who consolidates these models and makes the results accessible to the participating expert users. In the third scenario, the expert users participate in process modeling via video conferenc-

---

[303] See Dick, Duncanson (1999).
[304] See Chapter 2.3.

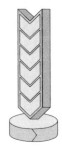

ing which, in this case, is coordinated by a centralized method expert. The results of modeling, in turn, are accessible by the expert users.

10.7.2
Scenario 1: Modeling on centralized server by decentralized clients

In the first scenario, method experts working at decentralized places can access a centralized server via the Internet in order to work with the process models stored on this server. The method experts use TCP/IP-based modeling clients (e.g., ARIS-Toolset) as well as Java Applets to work on a remote computer (in this case the modeling server) using a web-browser. In this scenario, changes to the process models become immediately visible to all process participants. The simultaneous access of two method experts to one process model, however, can cause problems for reasons of consistency. Reliability and speed of an Internet connection are other essential factors for the success of this scenario. A break in a connection during synchronous modeling may lead to inconsistent conditions in the modeling database if model changes are saved immediately in the database while the computer of the related participant is disconnected from the centralized server before the operation is confirmed. However, suitable technical actions, such as the use of persistent message queues (e. g., IBM MQSeries), are able to reduce this risk. The decentralized participants in this scenario must have an in-depth knowledge of the modeling method since they work directly with this tool. This widely restricts modeling by expert users who are untrained in the application of methods. Therefore, of greatest advantage is a fast and, above all, an asynchronous processing of process models by a group of method experts. If the affected company wants to model business processes, for example, which run between Germany and the United States, the time difference may favor a Continuous Process Engineering. When the method experts in Germany have finished their work, the method experts in the States pick up the appropriate changes and continue working.

TCP/IP capable clients

Knowledge of modeling method is required

Continuous Process Engineering

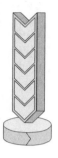

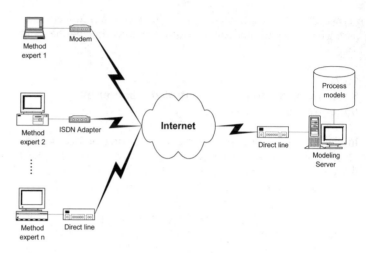

Fig. 10.15. Decentralized modeling via Internet

10.7.3
Scenario 2: Decentralized modeling and consolidation through moderation

No real-time access required

Moderator consolidates the models...

...and eliminates modeling errors

While the models were accessed in real time in the first scenario, the process models in the second scenario are received by a centralized moderator who consolidates them. The method experts can call up these consolidated models via www (see Chapter 8.5) or via other Internet services. The models are transferred to the moderator who checks them for correct syntax and consistent semantics. In this case also, knowledge of the modeling method is required by the decentralized units, but the moderator is, nevertheless, entitled to correct detected violations of modeling guidelines. Through the bundled transfer of process models via e-mail or FTP, this procedure is particularly suitable in those cases where high communication costs from decentralized units who are permanently connected to the Internet becomes uneconomical. Here, the models can be generated offline and then transferred to the moderator as a package.

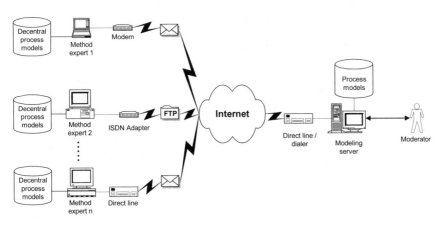
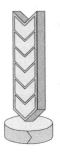

Fig. 10.16. Decentralized modeling with centralized moderation

10.7.4
Scenario 3: Centralized modeling with participation of decentralized expert users

In the third scenario, expert users participate in a virtual meeting (e. g., Microsoft NetMeeting or Intel ProShare) using audio and / or videoconference technology. The participants in this conference communicate via video images and / or audio connections. In this case, the generation of process models is the task of a centralized method expert (or team of method experts) who can be the moderator of the discussion as well. This function, however, can also be taken over by a second person. The modeling results are made accessible to the participants via a www-site or via a site that can be accessed in this videoconference. Technologically, this can be done, for example, by releasing modeling applications in a Microsoft NetMeeting. Here, a centralized moderator is not absolutely required. Single expert users can change a process model slightly in cooperation with the method expert. This model is then either transferred to a centralized moderator later, or entered directly into the model database by the decentralized method expert.

Video- or Audio conference

Application sharing

Decentralized comparison, centralized consolidation

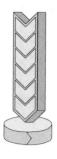

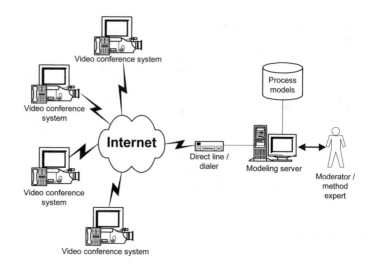

Fig. 10.17. Centralized modeling using video conference technology

*No method
knowledge
required*

In this scenario, the expert users do not need to know the modeling method in detail since the method experts are responsible for the consistency and syntactic correctness of the models. Centralized modeling makes the models become immediately visible to the expert users so that smaller corrections can be made all at once.

10.7.5
Protection of Internet-based modeling methods

*Risks of open
networks*

*Encryption
mechanisms*

What all the mentioned procedures have in common is that the security of data transfer via the Internet is lower than data transfer within a closed company network. Therefore, security mechanisms must be implemented that prevent unauthorized access to the transferred data. If the process models are accessed in real time, the transferred data can be encoded. Popular encoding mechanisms are, for example, DES (Data Encryption Standard), IDEA (International Data Encryption Algorithm), PGP (Pretty Good Privacy), RSA (Rivest, Shamir, Adleman) as well as SSL (Secure Socket Layer).[305] These procedures differ in the encryption algorithms as well as in the lengths of the related keys. In general, a longer key is more difficult to decode than a shorter key. The greater the secu-

---

[305] For encoding mechanisms, see, for example, Wobst (1998); Schneier (1994) and Davis, Price (1989).

rity requirements, however, the longer the computing time, which, in particular with modeling in real time, may lead to bottlenecks.

Another security aspect is the access of external computers to data in the company-owned Intranet. Here, it must be ensured that only authorized people are allowed to access the stored data, i.e., an authentication is required. An authentication uniquely identifies the inquiring person. Normally, authentication is done by entering a password that is known only to the inquiring person (authentication by knowledge). If greater security mechanisms are required, the authentication by knowledge and possession (e. g., entry of a password which is verified based on the contents of the chip-card of the user) and / or the authentication by biometric features (e. g., reading a fingerprint) are conceivable possibilities to restrict access from the outside to a correspondingly protected network.

External access to the Intranet

Additional security mechanisms allow only dedicated computers to access the Intranet (e.g., by maintaining a related table of IP-addresses) and / or to restrict access to the local network via hardware and software components, so-called "firewalls".

Additional security mechanisms

10.7.6
Evaluation of Internet-based process modeling

Internet-based process modeling offers companies the opportunity to collect the distributed knowledge of the user departments without the time and costs associated with centralized modeling teams and to integrate this knowledge into the process models. Here, it can be distinguished between moderated and non-moderated, as well as between synchronous (i.e., real time based) and asynchronous modeling scenarios. Depending on the technology, the participating individuals need to have more or less detailed knowledge of the modeling method. Less detailed knowledge is required with modern technologies since here the moderator can take corrective action, while in synchronous scenarios without a moderator, the model database is accessed directly. Criteria for the selection of alternatives are the existing technical infrastructure of the company, the scope of the modeling project, the size and distribution of the project teams, as well as the available knowledge of the modeling method of the participating persons. Finally, Table 10.1 compares the described scenarios.

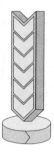

Table 10.1. Summary of scenarios

Technology	Type of Access	Consolidation of Models	Required Knowledge of Methods
Internet-Clients with centralized Server	Synchronous	by method experts	High
Internet-Clients with Mail / FTP-Access	Asynchronous	by moderator	Medium
Audio / Video conference	Synchronous	by moderator / method experts	Low

All three possibilities need to be prepared as described in Chapter 3, i.e., the modeling target must be determined, a business process framework must be developed, the modeling conventions have to be defined, and the modeling width and depth with hierarchical levels and the modeling granularity must be determined. Centralized modeling conventions are of major importance, in particular with decentralized modeling where the communication between the method experts (in spite of video conferencing and www-options) is often less intensive than with centralized modeling teams.

Appendix: Modeling Conventions

This appendix contains selected extracts from the modeling standard that was used in the DeTe Immobilien project. It concretizes the explanations given in Chapter 3. The examples refer primarily to the modeling of process models with event-controlled process chains and / or value chains as well as to the modeling of terminology.

A.1
General conventions

A.1.1
Business Process Framework

Table A.1 describes the roles involved in modeling.

Table A.1. Business Process Framework

Role	Tasks / Responsibilities	Job / Organization
Server manager (User "dbAdmin")	Administration of ARIS-Toolset (Installation of new modules and releases, reorganization)	• DTI-Organization
Database manager (User "System")	Database definition, setup of user / user groups / assignment of rights, definition of model filters and reports, creation and maintenance of group structure	• DTI-Organization
Technical tester	Check for observance of conventions in this document, maintaining of consistency with other models, use of specification copies, adaptation and further development of modeling standards, maintenance of module ARIS-configuration, in particular consistency checks	• DTI-Organization • University of Münster

PSC	Final release of top model level	• Project Steering Committee
User Represent-atives	Responsible for correct contents of the models, initialization of model revision, availability of interview partners for the modeling experts, promotion of model acceptance in the user departments, distribution of models in the user departments	• User Represen-tatives
Modeling experts	Generation of an information model according to the specifications of this document.	• DTI-Organization • University of Münster
Process Godfa-ther	Responsible for correct contents of the models, initialization of model revisions, availability of interview partners for the modeling experts, promotion of model acceptance in the user departments	• The known process god-fathers

A.1.2
Notes to the documentation of standards

The following abbreviations are used in the documentation that describe model-, object-, and edge type as well as the edge attribute position (refer to Table A.2).

Table A.2. Explanation of abbreviations used in this document

Abbreviation	Explanation
M	Has to be specified, i.e., a perspective-wide obligation exists for **maintenance**.
R	Has to be specified when the described **rule** is true. This obligation results from dependence on the relevance of the individual perspectives. A restriction, for example, is given when the perspective that is required in this field is irrelevant for a model (e.g., no maintenance of frequency of events with process models that are not subject to simulation). A perspective-wide example is the obligation to maintain plain texts if abbreviations are used.
K	May be specified, the specification has to follow the described **conventions**.

A.2
Special conventions

A.2.1
Conventional object type

A.2.1.1
Generally valid attribute types for every object type

The following attribute conventions are generally valid for every object type. The following explanations of the individual object types refer to either concretizing or amendment of these conventions.

Table A.3. Generally valid attribute types for every object type

Object Attribute Type	Convention	
Identifier	Identifier is automatically assigned by the ARIS-Toolset.	m
Plain text	Plain text, if an abbreviation has been specified in this field. This plain text must be consistent with the abbreviation list which will become an integral part of the glossary, and will be uniform in all databases	r
Note / Example	Comment, being a detailed explanation of a definition (without further restrictions), Example beginning with "Ex.:"	k
Processing flag	To be used for object types which still must be clarified (need to clarify = "?"), the models presented for release must not have any specifications.	k
Author	Name and position have to be entered, this field is filled by inheritance	m
Source and / or source indication / source	Origin of information for "Name" plus definition and eventual designation, version, and date when this information was created	k
Short name	Short name if a company-wide valid short name exists. The short name starts with a capital letter	r
System attribute / External 1	Link to Winword-documents that contain detailed explanations of the object. If required, references to additional project documents can be inserted.	k

Text	Notes, comments, open questions, user requirements, etc. This attribute is not available in all object types.	k

A.2.2
Object types to be used

A.2.2.1
Application system

 Carrier if DP functions will support user functions.
Definition of types for applications and application versions

Table A.4. Attributes of application system

Object Attribute Type	Convention	Example	
Name	For applications: Short name of application system, first character must be a capital letter. For application versions: Short name of application system + line wrap + "V" + Version-, Release number etc.	• PRINS • SAP R/3 FI V 4.0b	m
Description / Definition	For Applications: Informal short description of functional scopes. For application versions: Informal short description of functional change of version.	• Quotation- and contract processing for accounts	m
Installation status / in use since	When the application version is already active, the date (format: MM/DD or MM/YY) of productive start has to be entered here.	• "05/09/01" or "05/01"	r
Installation status / realized up to	When the application version is not yet productive, the planned date (format: MM/DD or MM/YY) of productive start has to be entered here	• "05/02/00" or "05/00"	r

A.2.2.2
Event

Event

These conventions distinguish between business economic events, time events, and ad-hoc events.

- Business economic event:
  Business economic relevant status of process object, this status could have occurred earlier than expected because a function has been executed earlier or because of an impact from outside.
- Time event:
  Occurrence of a point of time that triggers a process.
- Ad-hoc event:
  Unforeseen event outside the normal process which is relevant for business economics.

In addition, there is a distinction between initialization events (e.g. "Invoice to be posted") and execution events ("Invoice is posted").

Table A.5. Attributes of event

Object Attribute Type	Convention	Example	
Name	Business economic and ad-hoc events are described in format "Process object + is + verb written in past participle". The process object must be defined as a technical term (see A.2.2.3) or be created.	• Order request is entered	m
	The orientation toward the related verb (name of function) is done in a pure sequence. The name of the event, in this case, should be consistent and complete in comparison with the verb (name of function).	• Post voucher → voucher is posted	
	Time events are named in the following format: "Point of time has occurred".	Start of month occurred	
	Another name convention is the corresponding name of the opposite event.	• Credit limit is surpassed, and credit limit is not surpassed	
	Initialization events such as "Invoice to be posted" should be avoided, if possible.		
	Should a full designation turn out to be impossible, the name can be cut at the end if the full name is registered under "Plain text"		
Plain text	Should it not be possible to enter a full name in attribute "Name", then the full name appears here.	• Execution plan is entered	r

In addition to functions, events are central object types of event-driven process chains.

A.2.2.3
Technical term

Term which needs to be defined for activities in conformity with the related target. This term is characterized by the fact that it designates a structured subject.

Table A.6. Attributes of technical term

Object Attribute Type	Convention	Example	
Name	Full designation in singular; first character has to be a capital letter. If a full name is not possible, the name can be shortened at the end if the full name is entered in attribute type "Plain text". If the name contains an adjective, this adjective must be separated from the substantive by a comma. When the technical term relates to a group of information objects, the name can be used in plural.	• Creditor • Business partner hierarchy type, business partner category • Business partner category, creditor • Personnel master data	m
Synonyms	Designation of synonymous technical terms. Synonyms must be stored as own technical term and be linked with "relates to" with specification "S" in edge attribute type "Edge Role".	• Supplier	k
Plain text	Should it not be possible to enter a full designation in attribute "Name", then the full designation appears here.	• Business partner hierarchy type Business partner category assignment	r

Description / Definition	Definition of technical term in the present. The semantics must be selected to result in the answer to the question "What is a xy?" Recommendation: "The technical term xy is a(n)" should be assumed intellectually. The original text is adopted from laws or DIN-standards. Application systems, media, and organizational units must not be used in the definition. Subtypes of an "is a"-relation requires the supertype to be repeated.	• Example "Creditor": Business partner, for whom an obligation exists from a rendered service or delivery. • Example: Definition of subtype "Personnel creditor" with "is a" relation to supertype "Creditor" who receives wage or salary for his work.	m

Technical terms are used in the terminology model (technical concept "data view") and in event-driven process chains. In event-driven process chains, the technical terms are in a "is input to" relation to the functions.

A.2.2.4
Function

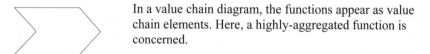

A function represents a task that is executed on one or multiple process objects.

The process guide serves as an interface symbol to refer to preceding or succeeding process models. It can only be positioned at the start or at the end of a process model and can only be linked to events.

In a value chain diagram, the functions appear as value chain elements. Here, a highly-aggregated function is concerned.

Table A.7. Attribute of function

Object Attribute Type	Convention	Example	
Name	Composed term of process object (substantive in singular) and action (verb in present infinitive). The process object must have been defined already as a technical term (A.2.2.3) or has to be created. In a deviation from this rule, in particular in value chain diagrams, highly aggregated functions may only consist of one substantive (in singular). If a full designation is not possible, the end of the name can be shortened. In this case, however, an entry of the full name is required in attribute type "Plain text".	• Compare Plan with Turnover • Customer Service • Facility Management	m
Plain text	If it is not possible to store a full designation in attribute "Name", the full designation is stored here.	• Facility Development	r
Description / Definition	Application systems, highly aggregated functions, organiza-tional units and media must not be used in the definition. When using technical terms, these technical terms must already exist or have to be created.	• Facility management includes the site and project development as well as their realization	k

Besides events, functions are the central object types within event-driven process chains. In value chain diagrams and in function assignment diagrams, functions are the sole central object types.

 When hierarching functions in an event-driven process chain, it must be observed that the start- and end events of the function match the detailed models (specification copy of events and linking rules). The name of the detailed model corresponds to the name of the hierarchical function (the name is automatically adopted by the ARIS toolset when the function is hierarchied). When using process guides, it has to be observed that the start- and end events of the process guide match with those in the process model to which the process guide refers.

In the following, some name conventions for dedicated functions are described in the form of an action catalog. This catalog has to be up-dated continuously.

Name conventions for functions (action catalog)

Table A.8. Action catalog

Enter	Formalizing an object in data and making it available to the viewed area (e.g.: manual entry in a DP-system, to enter data in a form in writing).
Calculate	Determination of an object based on existing data and calculation rules.
Inform	Availability of data of an object for personnel resources.
Verify	Verification of features of an object based on defined verification rules.

A.2.2.5
Job / Position

Personnel resource whose scope of tasks corresponds exactly to a person. Several employees who all execute the same tasks can occupy a job.

Table A.9. Attributes of job

Object Attribute Type	Convention	Example	
Name	Short name of task carrier and / or job.	• RsL Pe • RsLPe Mannheim	m
Short name	Abbreviation of job name when a task carrier exists	• NLL GK	r
Description / Definition	Informal short description of functional scope, official job description, if available		k
Plain text	Plain text of task carrier when a task carrier exists	• Key Account Manager	r

Object type "person" is used in event-driven process chains and – in case of a high degree of model detailing – also in organizational charts.

A.2.3
Conventions model type

A.2.3.1
Generally valid attribute types for every model type

The model attributes are maintained in the Model Manager. They appear, in part, on the printout in the header or footer.

Table A.10. Generally valid attribute types (in model header)

Model Attribute Type	Convention	Example	
Plain text	Plain text of model name if no full name could be entered in attribute "Name".	• Performance process of technical network division	r
Description / Definition	Model type according to guide. Can be directly copied from the related model conventions	• Function tree	m
Note, Example	Core process- or support process, external process (at customer).	• Core process	
Model status / Status	Specification of a model still under development	• in process	r
Model status / since / on	Start date or end date of processing (= release for verification) (Format: MMDDYYYY).	• 02/22/2001	m
Author	Name of modeling expert and name of project / job.	• Mr. Müller, Process-modeling	m
Certification / Created on	Date of release for verification of subject (Format MMDDYYYY).	• 04/19/2001	r
Certification / Verified on	Date of positive result (certification) of verified subject (Format MMDDYYYY).	• 05/02/2001	r
Certification / Verified by	Name of responsible person who verifies the subject and name of project / job.	• Ms. Kaiser, Controling	r
Certification / Released on	Date of release of process model by process responsible	• Ms. Kaiser, Controling	r
Certification / Released by	Name of process responsible and name of project / job	• Mr. Stötz, Accounting	r

A.2.4
Model types to be used

A.2.4.1
Event-driven process chain

Valid model attribute types

Table A.11. Valid model attribute types of EPC (Event-driven Process Chain)

Model Attribute Type	Convention	Example	
Name	Is automatically proposed by ARIS when the hierarching is done from a super-ordinated function. Composed term of process object (substantive in singular) and action (Verb in present infinitive). In a deviation from this rule, the names of top processes may consist of one substantive, only (in singular)	• Distribute mail • Sales	m
Definition	Designation of degree of detailing of a process, according to project-specific convention; core- or support process.	Core process	m

Valid object types

Table A.12. Valid object types of EPC (Event-driven Process Chain)

Object Type	Symbol	Object Attribute Type; -position (No.)	
Application system type	Application System Type	• Name (middle)	m
Event	Event	• Name (middle)	m
Technical term	Technical term FB	• Name (middle)	m
Function	Function Process Guide	• Name (middle)	m
			m
Organizational unit	Organizational Unit	• Name (middle)	m
Business partner	External Person ext	• Name (middle)	m
Rule	∧ ∨ XOR	• Name (middle)	m
Job	Job	• Name (middle)	m

Valid relationship type records

Table A.13. Valid relationship type records of EPC (Event-driven Process Chain)

Source Object Type	Edge Type	Target Object Type
Technical term	is input for …	Function
Function	changes ...	Technical term
Function	archives ...	Technical term
Function	creates ...	Technical term
Function	deletes ...	Technical term
Function	distributes ...	Technical term
Organizational unit, Job	is responsible for this subject	Function
Organizational unit, Job	is responsible for EDP	Function
Organizational unit, Job	executes ...	Function
Organizational unit, Job	has to be informed about …	Function
Organizational unit, Job	is involved in ...	Function
Organizational unit, Job	acts as consultant for ...	Function

Information objects of organizational structure are collected according to the "workflow principle", i.e., the assignment has to be as precise as possible. This can lead, for example, to a coexistence of organizational units and jobs alongside the process models.

In nominal modeling, the assignment of organizational working models and application system types to functions must only be done when this will not lead to a toorestricted process view.

Graphical convention – Arrangement of data / resources for a function (Function assignment diagram)

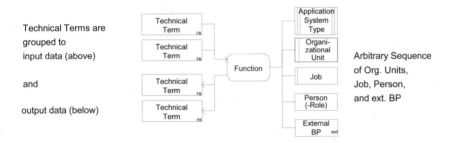

Fig. A.1. Graphical convention – function assignment diagram

The EPC (Event-driven Process Chain) is graphically arranged in such a way that the runtime frequency decreases from left to right – if possible.

Graphical conventions – Event-driven process chain

The process model starts with a minimum of one event and ends accordingly. Process guides, if any, can precede these events and / or follow them.

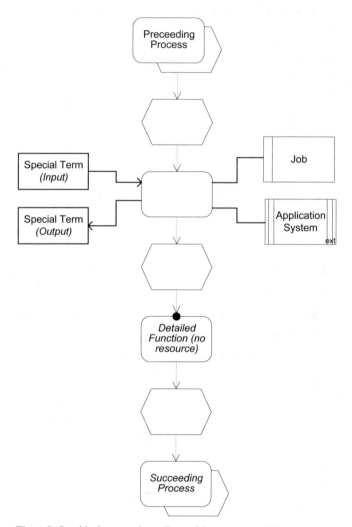

Fig. A.2. Graphical convention – Event-driven process chain

A.2.4.2
Terminology model

The terminology model serves as the basis for the glossary to be created. It contains general vocabulary of modeling (e.g., function, process) as well as DeTe Immobilien-specific vocabulary (e.g., facility, customer service). Glossaries are generated from the terminology model in Winword-, Windows-Helpfile- and HTML-format.

Valid model attribute types

Table A.14. Valid model attribute types of terminology model

Model Attribute Type	Convention Model Attribute Type	Example	
Name	Name of technical term of interest	• Job	m
Definition	Explanation of term. If definitions of third parties (e.g., DIN) are used, these sources have to be specified	• HR resource, whose scope of tasks relates to a person.	

Valid object types

Table A.15. Valid object types of terminology model

Object Type	Symbol	Object Attribute Type Position (No.);		Example Object Attribute Type, -position
Technical term	Technical Term FB	• Name (middle) • Short name (middle, left)	m k	Technical Term FB

Valid relationship type records

Table A.16. Valid relationship type records of terminology model

Source Object Type	Edge Type	Target Object Type
Technical term	relates to	Technical term
Technical term	includes	Technical term
Technical term	is a	Technical term
Technical term	classifies	Technical term
Technical term	is a feature of	Technical term
Technical term	is a copy of	Technical term

Functions and events from event-driven process chains should refer to technical terms, if technical terms are involved.

Concept for the maintenance of technical terms

In the following, it is explained how the (integrated) terminology model can be further maintained in the nominal modeling phase. The terminology is distributed internally within the company in Winword format and as HTML-file for agreement. The related proposal is addressed to all members of the project team whose task is to identify technical terms when generating models.

Every project member should concentrate on the existing terminology model. This, in particular, includes the study of DeTe Immobilien-internal vocabulary. Since, however, the mastering of all defined terms seems to be impossible due to the great number of technical terms, a "Model-Quality-Assurance" verifies to what degree process models and terminology models are consistent with each other. Process modeling and terminology modeling mutually influence one another. Therefore, a cooperation between modeling experts and the terminology model responsible is indispensable.

In nominal process modeling, new technical terms have to be identified in the generated value chains and event-driven process chains.

In those cases, where a new technical term is "obviously" being used, the modeling experts should immediately ask for a definition that is documented by the person who is responsible for the related log file. These new technical terms must be created and stored and include a definition and an indication of the author and source in their own terminology models(!), i.e., not as event-driven process chains within the groups Root\Technical terms\New Technical terms. The model name should clearly refer to the source (e.g., "Technical terms of Sales"). As far as possible, semantic relations (e.g., "is a feature of") should also be added, which is not possible within an EPC (Event-driven Process Chain).

The models of subgroup \New Technical terms are continuously emptied by the terminology model responsible, i.e., the terms are integrated into the terminology model after critical evaluation of the proposed definition and verification of the integrity of attribute maintenance, and then deleted, unless technical terms are concerned (after a note to the creator) and / or "returned" with the request for a more-detailed definition.

In addition, the group Technical terms contains two additional subgroups:

Group \Agreed Technical terms contains terminology models that were already approved by the user representatives. Group \Technical terms to be Agreed contains models whose acceptance is still pending.

After the first generation of nominal models, the relevant technical terms are compared with the existing terminology (in the ARIS-Toolset, agreed and not yet agreed models). If it turns out that new technical terms must be defined, the related definitions should be agreed upon in the first reconciliation process, if possible. If required, the interview partners as well can receive a list of technical terms to be defined and ask for definitions. These definitions could be sent, for example, by fax some days later.

In addition, the model quality assurance verifies to what extent the process models contain terms that are not yet defined in the terminology model. In this case, suitable representatives from the user departments are identified in agreement with the "modeling experts" who can be asked for term definitions. The aforementioned explanation of how to create new technical terms is applicable as well.

The terminology model is updated periodically. The special-term responsible will inform all project members continuously about any new vocabulary that is added to the terminology model.

Graphical conventions

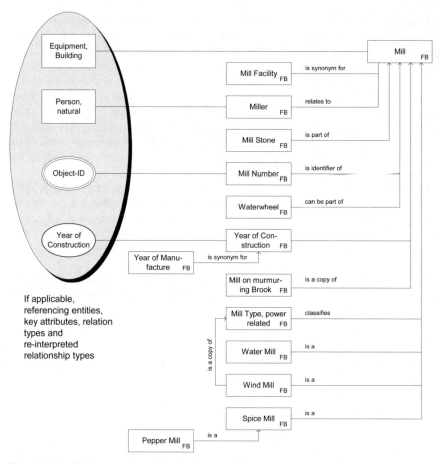

Fig. A.3. Graphical convention – Terminology model

A.2.4.3
Value chain diagram

Valid model attribute types

Table A.17. Valid model attribute types of value chain diagram

Model Attribute Type	Convention Model Attribute Type	Example	
Name	Is automatically proposed by ARIS toolset when hierarching from a super-ordinated function. Composed term of process object (substantive in singular) and action (verb in present infinitive). In a deviation from this rule, the names of top processes may consist of one substantive (in singular) only.	• Sales	m
Definition	Designation of the related level of detailing of the process according to project-specific convention; core- or support process.	• Core process	m

Valid object types

Table A.18. Valid object types of value chain diagram

Object Type	Symbol	Object Attribute Type; - position
Function		• Name (middle)
Organizational unit	Organizational Unit	• Name (middle)
Technical term	Technical Term FB	• Name (middle)

Valid relationship type records

Table A.19. Valid relationship type records of value chain diagram

Source Object Type	Edge Type	Target Object Type
Technical term	is input for	Function
Function	has output	Technical term
Function	changes	Technical term
Function	archives	Technical term
Function	creates	Technical term
Function	deletes	Technical term
Function	distributes	Technical term
Function	is super-ordinated in process orientation	Function
Function	is predecessor of	Function
Organizational unit	executes	Function
Organizational unit	is subject-responsible	Function
Organizational unit	is EDP responsible	Function
Organizational unit	has to informed about	Function
Organizational unit	is involved in	Function
Organizational unit	acts as consultant for	Function

Graphical Conventions

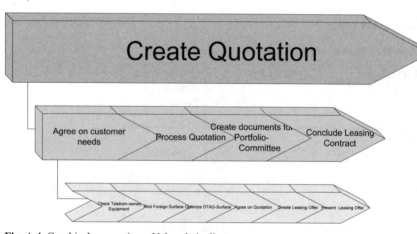

Fig. A.4. Graphical convention – Value chain diagram

Bibliography

Abell, D. F.: Defining the Business. The Starting Point of Strategic Planning. New York, NY 1980.

Adam, D.: Planung und Entscheidung. Modelle – Ziele – Methoden. 4th edition, Wiesbaden 1996.

Adam, D.: Produktions-Management. 9th edition, Wiesbaden 1998.

Adam, D.; Johannwille, U.: Die Komplexitätsfalle. In: Adam, D. (Ed.), Komplexitätsmanagement. SzU, Volume 61, Wiesbaden 1998, pp. 5-28.

Adam, D.; Rollberg, R.: Komplexitätskosten. In: DBW, 55, 1995, 5, pp. 667-670.

Al-Ani, A.: Continuous Improvement als Ergänzung des Business Reengineering. In: zfo, 65, 1996, 3, pp. 142-148.

Albach, H.: Maßstäbe für den Unternehmenserfolg. In: Hentzler, H. A. (Ed.): Handbuch Strategische Führung, Wiesbaden 1988, pp. 69-83.

Andrews, K. R.: The Concept of Corporate Strategy. 2nd edition, Homewood, IL 1987.

Arbeitskreis 'Organisation' der Schmalenbach-Gesellschaft. Deutsche Gesellschaft für Betriebswirtschaft e. V.: Organisation im Umbruch. In: zfbf, 48, 1996, 6, pp. 621-665.

Baan Business Innovation: Baan IVc Reference Models. Ede 1997.

Balzert, H.: Lehrbuch der Software-Technik. Software-Entwicklung. Heidelberg et al. 1996.

Barney, J. B.: Firm Resources and Sustained Competetive Advantage, In: Journal of Management, 17, 1991, pp. 99-120.

Batini, C.; Furlani, L.; Nardelli, E.: What is a good diagram? In: Chen, P. P.-S. (Ed.): Entity-Relationship Approach: The Use of ER Concept in Knowledge Representation. Proceedings of the 4th International Conference on the Entity-Relationship Approach, Elsevier, North Holland 1995, pp. 312-319.

Becker, J.; Ehlers, L.; Schütte, R.: Grundsätze ordnungsmäßiger Modellierung. In: Projektträger des BMBF beim DLR (Wolf, W.; Grote, U.) (Ed.): Tagungsband zur Statustagung des BMBF. Softwaretechnologie. Bonn 1998, pp. 63-93.

Becker, J.; Rosemann, M.; Schütte, R.: Grundsätze ordnungsmäßiger Modellierung (GoM), In: Wirtschaftsinformatik, 37, 1995, 5, pp. 435-445.

Becker, J.; Schütte, R.: Handelsinformationssysteme. Landsberg am Lech 1996.

Becker, J.; v. Uthmann, C.; zur Mühlen, M.; Rosemann, M.: Identifying the Workflow Potential of Business Processes. In: Sprague, R. (Ed.): Proceedings of the 32nd Hawaii International Conference on Systems Sciences (HICSS32), Wailea, HI 1999.

Becker, J.; zur Mühlen, M.: Rocks, Stones and Sand. Zur Granularität von Komponenten in Workflowmanagementsystemen. In: Information Management & Consulting, 13, 1999, 2, pp. 57-67.

A

B

Berg, C. C.: Organisationsgestaltung. Stuttgart et al. 1981.

Bleicher, K.: Das Konzept Integriertes Management. 4th edition, Frankfurt am Main, New York, NY 1996.

Boehm, B. W.: A Spiral Model of Software Development and Enhancement. In: IEEE Computer, 21, 1988, 6, pp. 61-72.

Bogaschewsky, R.; Rollberg, R.: Prozeßorientiertes Management. Berlin et al. 1998.

Bourgeois, L. J.; Brodwin, D. R.: Strategic Implementation. In: Strategic Management Journal, 1984, 5, pp. 241-264.

Brandenburg, F.; Jünger, M.; Mutzel, P.: Algorithmen zum automatischen Zeichnen von Graphen. In: Informatik-Spektrum, 20, 1997, 4, pp. 199-207.

Buresch, M.; Kirmair, M.; Cerny, A.: Auswahl von Organisations-Engineering-Tools. In: zfo, 66, 1997, 6, pp. 367-373.

Camp, R. C.: Benchmarking. The search for industry best practices that lead to superior performance, Milwaukee, WI 1989.

Champy, J.: Quantensprünge sind angesagt. In: Top Business, 11, 1994, pp. 86-94.

Chan, R.; Rosemann. M.: Integrating Knowledge into Process Models – Insights from Two Case Studies. In: Proceedings of the 6th Pacific Asia Conference on Information Systems (PACIS 2002). Tokyo 2002.

Chandler, A. D.: Strategy and Structure. Chapters in the History of Industrial Enterprise. Cambridge, MA, London 1962.

Chrobrok, R.; Tiemeyer, E.: Geschäftsprozeßorganisation. In: zfo, 65, 1996, 3, pp. 165-172.

Coyne, K. P.; Hall, St. J. D.; Gorman Clifford, P.: Is your core competence a mirage? In: The McKinsey Quarterly, 34, 1997, 1, pp. 40-54.

Curren, T.; Keller, G.: SAP R/3 Business Blueprint. Understanding the business process reference model. Upper Saddle River, NJ 1998.

Darke, P.; Shanks, G.: Stakeholder Viewpoints in Requirements Definition. In: Requirements Engineering, 1, 1996, 1, pp. 88-105.

Davenport, T. H.: Process innovation. Reengineering work through information technology, Boston, MA 1993.

Davis, D. W.; Price, W. L.: Security for Computer Networks. An Introduction to Data Security in Teleprocessing and Electronic Funds Transfer. New York, NY et al. 1989.

Dehnert, J.: Making EPCs fit for Workflow Management. In: Mitteilungen der GI-Fachgruppe Entwicklungsmethoden für Informationssysteme und deren Anwendung, 23, 2003, 1, pp. 12-26.

DGQ: Qualitätssicherungs-Handbuch und Verfahrensanweisungen. Ein Leitfaden für die Erstellung. Aufbau, Einführung, Musterbeispiele. DGQ Schrift Nr. 12-62, 2nd edition, Berlin 1991.

Dick, G.; Duncanson, I.: Telecommuting. In: Sprague, R. (Ed.): Proceedings of the 32nd Hawaii International Conference on Systems Sciences (HICSS32), Wailea, HI 1999.

Emrich, C.: Business Process Reengineering. In: io management, 65, 1996, 6, pp. 53-56.

Endl, R.; Fritz, B.: Integration von Standardsoftware in das unternehmensweite Datenmodell. In: Information Management, 7, 1992, 3, pp. 38-44.

Erdmann, T.: Modellbasierte Einführung von Oracle Applications. In: Maicher, M.; Scheruhn, H. J. (Ed.): Informationsmodellierung. Referenzmodelle und Werkzeuge, Wiesbaden 1998, pp. 253-274.

Esswein, W.: Das Rollenmodell der Organisation. In: Wirtschaftsinformatik, 35, 1993, 6, pp. 551-561.

C

D

E

Eversheim, W.: Prozeßorientierte Unternehmensorganisation. Konzepte und Methoden zur Gestaltung „schlanker" Organisationen. Berlin et al. 1995.

Fank, M.: Tools zur Geschäftsprozeßorganisation. Entscheidungskriterien, Fallstudienorientierung, Produktvergleiche. Braunschweig, Wiesbaden 1998.

Finkeißen, A.; Forschner, M.; Häge, M.: Werkzeuge zur Prozeßanalyse und -optimierung. In: Controlling, 8, 1996, 1, pp. 58-67.

Franz, K.-P.: Prozeßmanagement und Prozeßkostenrechnung. In: Schmalen-bach-Gesellschaft. Deutsche Gesellschaft für Betriebswirtschaft e. V. (Ed.): Reengineering. Konzepte und Umsetzung innovativer Strategien und Strukturen. Kongress-Dokumentation 48. Deutscher Betriebswirtschaftler-Tag 1994. Stuttgart 1995, pp. 117-126.

Frese, E.: Grundlagen der Organisation. 6th edition, Wiesbaden 1995.

Frese, E.; v. Werder, A.: Organisation als strategischer Wettbewerbsfaktor. In: Frese, E. (Ed.): Organisationsstrategien zur Sicherung der Wettbewerbs-fähigkeit. Lösungen deutscher Unternehmungen. Düsseldorf, Frankfurt am Main 1994, pp. 1-28.

Gaitanides, M.: Prozeßorganisation. Entwicklung, Ansätze und Programme prozeßorientierter Organisationsgestaltung. München 1983.

Gaitanides, M.; Scholz, R.; Vrohlings, A.; Raster, M.: Prozeßmanagement. Konzepte, Umsetzungen und Erfahrungen des Reeningeering. München, Wien 1994.

Gaitanides, M.; Sjurts, I.: Wettbewerbsvorteile durch Prozeßmanagement: Eine ressourcenorientierte Analyse. In: Corsten, H.; Will, T. (Ed.): Unternehmungsführung im Wandel. Strategien zur Sicherung des Erfolgspotentials. Stuttgart et al. 1995.

Gälweiler, A.: Strategische Unternehmensführung. 2nd edition, Frankfurt am Main, New York, NY 1990.

Galler, J.: Vom Geschäftsprozeßmodell zum Workflow-Modell. Wiesbaden 1997.

Gerard, P.: Unternehmensdaten-Modelle haben Erwartungen nicht erfüllt. In: Computerwoche, 42, 1993.

Girth, W.: Methoden und Techniken für Prozeßanalysen und Redesign. In: Krickl, O. G. (Ed.): Geschäftsprozeßmanagement. Heidelberg 1994.

Gomez, P.: Neue Trends in der Konzernorganisation. In: zfo, 61, 1992, 3, pp. 166-172.

Gutenberg, E.: Betriebswirtschaftslehre. Volume I: Die Produktion. 24th edition, Berlin et al. 1983.

Hagel III, J.: Fallacies in organizing for performance. In: The McKinsey Quarterly, 31, 1994, 2, pp. 97-106.

Hagemeyer, J.; Rolles, R.: Aus Informationsmodellen weltweit verfügbares Wissen machen. In: Information Management & Consulting, 11, 1997, Sonderausgabe Business Engineering, pp. 56-59.

Hahn, D.: PuK, Controllingkonzepte. Planung und Kontrolle, Planungs- und Kontrollsysteme, Planungs- und Kontrollrechnung. 5th edition, Wiesbaden 1996.

Hammer, M.: Beyond Reengineering. New York, NY 1996.

Hammer, M.; Champy, J.: Reengineering the Corporation. A Manifesto for Business Revolution, New York, NY 1993.

Hars, A.: Referenzmodelle. Grundlagen effizienter Datenmodellierung. Wiesbaden 1994.

Hedley, B.: Strategy and the „business portfolio". In: Long Range Planning, 10, 1977, 1.

F

G

H

Heib, R.: Business Process Reengineering mit ARIS-Modellen. In: Scheer, A.-W. (Ed.): ARIS – Vom Geschäftsprozeß zum Anwendungssystem. 3rd edition, Berlin et al. 1998, pp. 147-153.

Helling, K.: ISO 9000-Zertifizierung mit ARIS-Modellen. In: Scheer, A.-W. (Ed.): ARIS – Vom Geschäftsprozeß zum Anwendungssystem. 3rd edition, Berlin et al. 1998.

Hess, T.; Brecht, H.: State of the art des Business process redesign. Wiesbaden 1996.

Hinterhuber, H. H.: Strategische Unternehmensführung. Bd. 1: Strategisches Denken. Berlin, New York, NY 1992.

Hoffmann, W.; Kirsch, J.; Scheer, A.-W.: Modellierung mit Ereignisgesteuerten Prozeßketten. Methodenhandbuch, Stand: Dezember 1992. Veröffentlichungen des Instituts für Wirtschaftsinformatik, Volume 101, Saarbrücken 1993.

Holsapple, C. W.; Joshi, K. D.: Description and Analysis of Existing Knowledge Management Frameworks. In: Sprague, R. (Ed.): Proceedings of the 32nd Hawaii International Conference on Systems Sciences (HICSS32), Wailea, HI 1999.

Horváth, P.: Die „Vorderseite" der Prozeßorientierung. In: Controlling, 9, 1997, 2, pp. 114.

Horváth, P.; Gleich, R.: Prozeß-Benchmarking in der Maschinenbaubranche. In: ZwF, 93, 1998, 7-8, pp. 325-329.

IBM MQSeries Workflow: Getting Started with Buildtime Version 3.1.2. Document No. SH-12-6286-02. IBM Deutschland Entwicklungs GmbH, Böblingen 1998.

Ishikawa, K.: What is Total Quality Control? The Japanese Way, Englewood Cliffs, NJ 1985.

Keller, G; Partner: SAP R/3 prozeßorientiert anwenden. 3rd edition. Bonn et al. 1999.

Kieser, A.: Moden & Mythen des Organisierens. In: DBW, 56, 1996, 1, pp. 21-39.

Kieser, A.: Implementierungsmanagement im Zeichen von Moden und Mythen des Organisierens. In: Nippa, M.; Scharfenberg, H. (Ed.): Implementierungsmanagement. Über die Kunst, Reengineeringkonzepte erfolgreich umzusetzen. Wiesbaden 1997, pp. 81-102.

Kirchmer, M.: Business Process Oriented Implementation of Standard Software. How to Achieve Competitive Advantage Quickly and Efficiently. Berlin et al. 1998.

Kohl, U.; Schinn, G.: Dynamic Enterprise Modeling. In: Maicher, M.; Scheruhn, H. J. (Ed.): Informationsmodellierung. Referenzmodelle und Werkzeuge. Wiesbaden 1998, pp. 291-316.

Körmeier, K.: Prozeßorientierte Unternehmensgestaltung. In: WiSt, 24, 1995, 5, pp. 259-261.

Kosiol, E.: Die Unternehmung als wirtschaftliches Aktionszentrum. Reinbeck 1966.

Kosiol, E.: Aufbauorganisation. In: Grochla, E. (Ed.): Handwörterbuch der Organisation. Stuttgart 1969, Sp. 172-191.

Kosiol, E.: Organisation der Unternehmung. 2nd edition. Wiesbaden 1976.

Krahn, A.: Vom Prozeßmonitoring zum Prozeßmanagement. Ein Vorgehensmodell zur Indikatorenherleitung für ein Prozeß-Monitoring-System – dargestellt an der Firma H. Hoffmann-La Roche AG. Bern et al. 1998.

Krickl, O. G.: Business Redesign – Prozeßorientierte Organisationsgestaltung und Informationstechnologie. In: Krickl, O. G. (Ed.): Geschäftsprozeßmanagement, Heidelberg 1994, pp. 17-38.

Krüger, W. (1994a): Organisation der Unternehmung. 3rd edition, Stuttgart, Berlin, Köln 1994.

Krüger, W. (1994b): Umsetzung neuer Organisationsstrategien: Das Implementierungsproblem. In: Frese, E.; Maly, W. (Ed.): zfbf, 1994, Sonderheft 33, pp. 197-221.

Kugeler, M.; Rosemann, M.: Fachbegriffsmodellierung für betriebliche Informationssysteme und zur Unterstützung der Unternehmenskommunikation. In: Rundbriefe des GI-Fachausschusses, 5, 1998, 2, pp. 8-15.

Küting, K.; Lorson, P.: Benchmarking von Geschäftsprozessen als Instrument der Geschäftsprozeßanalyse. In: Berkau, C.; Hirschmann, P. (Ed.): Kostenorientiertes Geschäftsprozeßmanagement. Methoden, Werkzeuge, Erfahrungen. München 1996. pp. 121-140.

Lamla, J.: Prozeßbenchmarking. München 1995.

Lehmann, H.: Aufbauorganisation. In: Grochla, E.; Wittmann, W. (Ed.): Handwörterbuch der Betriebswirtschaftslehre. 4th edition, Stuttgart 1974, Sp. 290-298.

Ließmann, H.; Engelhardt, A.: OAG (Open Applications Group). In: Wirtschaftsinformatik, 40, 1998, 1, pp. 73-75.

Lindland, O. I.; Sindre, G.; Sølvberg, A.: Understanding Quality in Conceptual Modeling. In: IEEE Software, 11, 1994, 2, pp. 42-49.

Lohse, J. M.: Neue Beratungsanforderungen bei der Implementierung von Reengineeringkonzepten. In: Nippa, M.; Scharfenberg, H. (Ed.): Implementierungsmanagement. Über die Kunst, Reengineeringkonzepte erfolgreich umzusetzen. Wiesbaden 1997, pp. 189-200.

Maier, R.: Qualität von Datenmodellen. Wiesbaden 1996.

Meffert, H.: Marketing-Management. Analyse – Strategie – Implementierung. Wiesbaden 1994.

Meffert, H.: Marketing. Grundlagen Marktorientierter Unternehmensführung. 8th edition, Wiesbaden 1998.

Mertens, P.: Die Kehrseite der Prozeßorientierung. In: Controlling, 9, 1997, 2, pp. 110-111.

Moody, D. L.; Shanks, G.: What Makes a Good Data Model? In: Loucopoulos, P. (Ed.): Proceedings of the 13th International Conference on the Entity-Relationship Approach, Berlin et al. 1994, pp. 94-111.

Moody, D. L.; Shanks, G.: What Makes a Good Data Model? In: The Australian Computer Journal, 30, 1998, 3, pp. 97-110.

Müller-Merbach, H.: Operations Research. Methoden und Modelle der Optimalplanung. 3rd edition, München 1988.

Müller-Stewens, G.: Strategie und Organisationsstruktur. In: Frese, E. (Ed.): Handwörterbuch der Organisation. 3rd edition, Stuttgart 1992, Sp. 2344-2355.

Nonaka, I.; Takeuchi, H.: Die Organisation des Wissens. Wie japanische Unternehmen eine brachliegende Ressource nutzbar machen. Frankfurt am Main 1997.

Nordsieck, F.: Grundlagen der Organisationslehre, Stuttgart 1934.

Nordsieck, F.: Betriebsorganisation. Lehre und Technik, Textband. 2nd edition, Stuttgart 1972.

L

M

N

O

Oess, A.: Total Quality Management. Die ganzheitliche Qualitätsstrategie. Wiesbaden 1993.

Olfert, K.; Steinbuch, P. A.: Personalwirtschaft. 5th edition, Ludwigshafen (Rhein) 1993.

Ortner, E.: Methodenneutraler Fachentwurf. Stuttgart, Leipzig 1997.

Österle, H.; Brenner, W.: Integration durch Synonymerkennung. In: Information Management, 1, 1986, 2, pp. 54-62.

Osterloh, M.; Frost, J.: Prozeßmanagement als Kernkompetenz. Wie Sie Business Reengineering strategisch nutzen können. Wiesbaden 1998.

P

Porter, M. E.: Wettbewerbsvorteile. Spitzenleistungen erreichen und behaupten. Frankfurt am Main, New York, NY 1989.

Porter, M. E.: Wettbewerbsstrategie: Methoden zur Analyse von Branchen und Konkurrenten. 7th edition, Frankfurt am Main, New York, NY 1992.

Prahalad, C. K.; Hamel, G.: Competing for the Future. Boston, MA 1994.

Prahalad, C. K.; Hamel, G.: The Core Competence of the Corporation. In: Harvard Business Manager, 1990, May-June, pp. 79-91.

R

Raue, H.: Wiederverwendbare betriebliche Anwendungssysteme. Grundlagen und Methoden ihrer objektorientierten Entwicklung. Wiesbaden 1996.

Reiß, M.: Führungsaufgabe „Implementierung". In: Personal, 45, 1993, 12, pp. 551-559.

Reiß, M.: Reengineering. In: Horváth, P. (Ed.): Kunden und Prozesse im Fokus. Controlling und Reengineering, Stuttgart 1994, pp. 9-26.

Reiß, M.: Was ist schädlich an der Prozeßorientierung? In: Controlling, 9, 1997, 2, pp. 112-113.

Reitzenstein, N.: Durchgängige Prozeßmodellierung als Hilfsmittel für eine erfolgreiche Softwareimplementierung. In: Maicher, M.; Scheruhn, H. J. (Ed.): Informationsmodellierung. Referenzmodelle und Werkzeuge. Wiesbaden 1998, pp. 383-394.

Remme, M.: Konstruktion von Geschäftsprozessen. Ein modellgestützter Ansatz durch Montage generischer Prozeßpartikel. Wiesbaden 1997.

Rosemann, M. (1996a): Komplexitätsmanagement in Prozeßmodellen. Methodenspezifische Gestaltungsempfehlungen für die Informationsmodellierung. Wiesbaden 1996.

Rosemann, M. (1996b): Prozeß- vs. Ressourceneffizienz. In: Wirtschaftsinformatik, 38, 1996, 6, pp. 653-654.

Rosemann, M. (1996c): Multiperspektivische Informationsmodellierung auf der Basis der Grundsätze ordnungsmäßiger Modellierung. In: Management & Computer, 4, 1996, 4, pp. 219-226.

Rosemann, M. (1998a): Die Grundsätze ordnungsmäßiger Modellierung. Intention, Entwicklung, Architektur und Multiperspektivität. In: Maicher, M., Scheruhn, H. J. (Ed.): Informationsmodellierung. Referenzmodelle und Werkzeuge. Wiesbaden 1998, pp. 1-21.

Rosemann, M. (1998b): Managing the Complexity of Multiperspective Information Models using the Guidelines of Modeling, In: Fowler, D., Dawson, L. (Ed.): Proceedings of the 3rd Australian Conference on Requirements Engineering. Geelong 1998, pp. 101-118.

Rosemann, M. (1998c): Facetten der Wirtschaftlichkeit von Workflowmanagementsystemen. In: io management, 6, 1998, 9, pp. 44-50.

Rosemann, M.: Using Reference Models within the Enterprise Resource Planning Lifecycle. In: Australian Accounting Review, 10, 2000, 3, pp. 19-30.

Rosemann, M.: Integrated Process and Knowledge Management. Special Issue on Knowledge Management. B-HERT News. Julz 2002, pp. 24-26.

Rosemann, M.; Rotthowe, Th.: Der Lösungsbeitrag von Prozeßmodellen bei der Einführung von SAP R/3 im Rechnungswesen. In: HMD, 32, 1995, 182, pp. 8-25.

Rosemann, M.; Rotthowe, Th.; Schütte, R.: Referenzmodelle zur Auswahl und Einführung von Standardsoftware. In: Wenzel, P. (Ed.): Business Computing mit SAP R/3. Modellierung, Customizing und Anwendung betriebswirtschaftlich-integrierter Geschäftsprozesse. Braunschweig, Wiesbaden 1999, pp. 197-216.

Rosemann, M.; Schütte, R.: Multiperspektivische Referenzmodellierung. In: Becker, J.; Rosemann, M.; Schütte, R. (Ed.): Referenzmodellierung. State-of-the-Art und Entwicklungsperspektiven. Heidelberg 1999, pp. 22-44.

Rosemann, M.; zur Mühlen, M.: Evaluation of Workflow Management Systems – A Meta-Model Approach. In: Australian Journal of Information Systems, 6, 1998, 1, pp. 103-116.

Royce, W. W.: Managing the Development of Large Software Systems. In: Proceedings of IEEE WESCON, Los Alamitos, CA 1970.

Scheer, A.-W.: EDV-orientierte Betriebswirtschaftslehre. 4th edition, Berlin et al. 1990.

Scheer, A.-W. (1998a): Wirtschaftsinformatik. Referenzmodelle für industrielle Geschäftsprozesse, Studienausgabe. 2nd edition, Berlin et al. 1998.

Scheer, A.-W. (1998b): ARIS – Vom Geschäftsprozeß zum Anwendungssystem. 3rd edition, Berlin et al. 1998.

Scheer, A.-W. (1998c): ARIS – Modellierungsmethoden, Metamodelle, Anwendungen. 3rd edition, Berlin et al. 1998.

Scheer, A.-W.: „ARIS – House of Business Engineering": Konzept zur Beschreibung und Ausführung von Referenzmodellen. In: Becker, J; Rosemann, M.; Schütte, R. (Ed.) Referenzmodellierung. State-of-the-Art und Entwicklungsperspektiven. Heidelberg 1999, pp. 2-21.

Schmidt, G.: Methode und Techniken der Organisation. 8th edition, Gießen 1989.

Schneier, B.: Applied Cryptography: Protocols, Algorithms, and Source Code in C. New York, NY 1995.

Scholz, R.: Geschäftsprozeßoptimierung. Crossfunktionale Rationalisierung oder strukturelle Reorganisation. 2nd edition, Bergisch Gladbach, Köln 1993.

Scholz, R.; Vrohlings, A. (1994a): Prozeß-Redesign und kontinuierliche Prozeßverbesserung. In: Gaitanides, M.; Scholz, R.; Vrohlings, A.; Raster, M. (Ed.): Prozeßmanagement. Konzepte, Umsetzungen und Erfahrungen des Reengineering. München, Wien 1994, pp. 99-122.

Scholz, R.; Vrohlings, A. (1994b): Realisierung von Prozeßmanagement. In: Gaitanides, M.; Scholz, R.; Vrohlings, A.; Raster, M. (Ed.): Prozeßmanagement. Konzepte, Umsetzungen und Erfahrungen des Reengineering, München, Wien 1994, pp. 21-36.

Schreyögg, G.: Organisation. Grundlagen moderner Organisationsgestaltung mit Fallstudien. Wiesbaden 1996.

Schuh, G.; Katzy, B. R.; Dresse, pp.: Prozeßmanagement erfolgreich einführen. In: io management, 64, 1995, 12, pp. 64-67.

Schulte-Zurhausen, M.: Organisation. München 1995.

Schütte, R.: Grundsätze ordnungsmäßiger Referenzmodellierung. Wiesbaden 1998.

S

Schütte, R.: Wissen und Information. Antinomie oder Integration zweier Grundbegriffe der Wirtschaftsinformatik. In: Scheer, A.-W.; Rosemann, M.; Schütte, R. (Ed.): Integrationsmanagement. Arbeitsbericht Nr. 65 des Instituts für Wirtschaftsinformatik, Münster 1999, pp. 144-161.

Schwarzer, B.; Krcmar, H.: Grundlagen der Prozeßorientierung. Eine vergleichende Untersuchung in der Elektronik- und Pharmaindustrie. Wiesbaden 1995.

Schweitzer, M.: Ablauforganisation. In: Grochla, E.; Wittmann, W. (Ed.): Handwörterbuch der Betriebswirtschaftslehre. 4th edition, Stuttgart 1974, Sp. 1-8.

Schwickert, A. C.; Fischer, K.: Der Geschäftsprozeß als formaler Prozeß. Definition, Eigenschaften und Arten. Mainz 1996.

Shanks, G.: Conceptual Data Modelling. In: Australian Journal of Information Systems, 4, 1997, 2, pp. 63-73.

Sommerlatte, T.; Wedekind, E.: Leistungsprozesse und Organisationsstruktur. In: Arthur D. Little (Ed.): Management der Hochleistungsorganisation. 2nd edition, Wiesbaden 1991, pp. 23-41.

Spiegel, H.: Methodik zur Analyse und Dokumentation fachlicher Begriffswelten innerhalb des Unternehmens TELEKOM. Darmstadt 1993.

Stahlknecht, P.; Hasenkamp, U.: Einführung in die Wirtschaftsinformatik. 8th edition, Berlin et al. 1997.

Stalk, G.; Evans, P.; Shulman, L. E.: Competing on Capabilities. In: Harvard Business Review, 70, 1992, 3-4, pp. 57-69.

Steinmann, H.; Schreyögg, G.: Management. Grundlagen der Unternehmensführung. Konzepte – Funktionen – Fallstudien. 4th edition, Wiesbaden 1997.

Storey, V. C.: Understanding Semantic Relationships. In: VLDB Journal, 2, 1993, 2, pp. 455-488.

Striening, H. D.: Prozeß-Management. Versuch eines integrierten Konzeptes situationsadäquater Gestaltung von Verwaltungsprozessen. Frankfurt am Main et al. 1988.

Strohmayr, W.; Schwarzmaier, C.: Finanzdienstleistungen prozeßorientiert gestalten. In: Nippa, M.; Picot, A. (Ed.): Prozeßmanagement und Reengineering. Die Praxis im deutschsprachigen Raum. Frankfurt am Main, New York, NY 1995, pp. 258-271.

Tamassia, D.; Di Battisti, G.; Batini, C.: Automatic graph drawing and readibility of diagrams. In: IEEE Transactions on Systems, Man and Cybernetics, 18, 1988, 1, pp. 61-79.

Theuvsen, L.: Business Reengineering. In: zfbf, 48, 1996, 1, pp. 65-82.

Theuvsen, L.: Merkmale und Problemfelder aktueller Organisationskonzepte. In: Nippa, M.; Scharfenberg, H. (Ed.): Implementierungsmanagement. Über die Kunst, Reengineeringkonzepte erfolgreich umzusetzen. Wiesbaden 1997, pp. 103-131.

Thiele, M.: Kernkompetenzorientierte Unternehmensstrukturen. Ansätze zur Neugestaltung von Geschäftsbereichsorganisationen. Wiesbaden 1997.

Thompson, A. A.; Strickland, A. J.: Crafting & Implementing Strategy. 6th edition, Chicago, IL 1995.

Tuomi, I.: Data is More Than Knowledge. In: Sprague, R. (Ed.): Proceedings of the 32nd Hawaii International Conference on Systems Sciences (HICSS32) Wailea, HI 1999.

T

v. Uthmann, C.: Machen Ereignisgesteuerte Prozeßketten (EPK) Petrinetze für die Geschäftsprozeßmodellierung obsolet? In: EMISA FORUM – Mitteilungen der GI-Fachgruppe „Entwicklungsmethoden für Informationssysteme und deren Anwendung", 1998, 1, pp. 100-107.

van Es, R.: Dynamic Enterprise Innovation. Establishing Continuous Improvement in Business. 3rd edition, Ede 1998.

Weidner, W.; Freitag, G.: Organisation in der Unternehmung: Aufbau- und Ablauforganisation. Methoden und Techniken praktischer Organisationsarbeit. 5th edition, München, Wien 1996.

Welge, M. K.; Al-Laham, A.: Strategisches Management, Organisation. In: Frese, E. (Ed.): Handwörterbuch der Organisation. 3rd edition, Stuttgart 1992, Sp. 2355ff.

Welge, M. K.; Al-Laham, A.: Struktur in Strategieprozessen. Ergebnisse einer explorativen empirischen Studie. In: ZfB, 68, 1998, 8, pp. 871-898.

Wild, J.: Grundlagen der Unternehmensplanung. 4th edition, Opladen 1982.

Witte, E.: Ablauforganisation. In: Grochla, E. (Ed.): Handwörterbuch der Organisation. Stuttgart 1969, Sp. 20-30.

Wobst, R.: Abenteuer Kryptologie. Methoden, Risiken und Nutzen der Datenverschlüsselung. Bonn 1998.

Workflow Management Coalition: Terminology & Glossary. The Workflow Management Coalition Specification. Document Number WfMC-TC-1011. Document Status – Issue 3.0. Winchester 1999.

Zahn, E.: Produktion als Wettbewerbsfaktor, In: Corsten, H.: (Ed.): Handbuch Produktionsmanagement. Strategie – Führung – Technologie – Schnittstellen. Wiesbaden 1994, pp. 241-258.

zur Mühlen, M.: Integrationsperspektiven des Workflowmanagement. In: Scheer, A.-W.; Rosemann, M.; Schütte, R.: (Ed.): Integrationsmanagement. Arbeitsbericht Nr. 65 des Instituts für Wirtschafsinformatik, Münster 1999, pp. 94-106.

V

W

Z

Index

A

B

C

D

E

F

G

H

I

J

K

L

M

Q

R

S

T

V

W

Authors

Prof. Dr. Jörg Becker
Director of the Department of Information Systems at the University of Muenster

- MBA (1982) and PhD (1987) both from the University of Saarland, Study of Business Administration and Economics, University of Michigan, Ann Arbor, USA (1980-81), Lecturer and Assistant Professor, Department of Information Systems, University of Saarland (1982-90), Consultant of IDS Gesellschaft für Integrierte Datenverarbeitungssysteme GmbH (1987), Since 1990 Full Professor at the University of Münster. Research areas: information management, information modeling, data management, logistics, retail information systems, management consultant for IT-strategy, projects in industry-, service- and retail companies.
- Project work: project management.

E-Mail: becker@wi.uni-muenster.de

Dipl.-Kfm. Dieter Kahn
Former Managing Director, Administration and Finance, DeTe Immobilien

- Study of Business Economics, TU Berlin (graduation 1960). Field of activity: Accounting at Berliner Druck- und Kartonagenfabrik; Internationale Industrie- und Verwaltungs AG, Berlin; Granus Glasfabrik und Granus Vertrieb, Aachen; Felten & Guilleaume, Nordenham; Broadcast Television Systems GmbH, Darmstadt; DeTeBau, Frankfurt. Managing Director of DeTe Immobilien (1993-2000). Retired since 2001.
- Project work: permanent member of project steering committee.

E-Mail: dkahn@t-online.de

Dr. Martin Kugeler

Executive Assistant to the CFO, DirectGroup Bertelsmann

- MBA (1996) from the University of Münster. Lecturer at the Department of Information Systems, University of Münster (1996-2001). Guest lecturer at the Queensland University of Technology (QUT), Brisbane (Australia, 2000). Since 2001 at the DirectGroup Bertelsmann.
- Project work: as-is modeling, to-be modeling of core processes sales, development of organizational structures in „facility management and service", communication and implementation of organizational structure and processes as well as modeling of technical terms.

E-Mail: martin@kugeler.de

Dipl.-Ing. (FH) Michael Laske

Specialist for Process and Management Consulting, T-Systems International GmbH

- Training Telecommunication Tradesman (1983-86), Study of communication technology, FH Osnabrück (1990-94). Fernmeldeamt Münster (1986-88); Fields of activity: IT-Development, Process Modeling and Process Management, Deutsche Telekom AG (1994-96). Process and project management at DeTe Immobilien. Since 2000 at T-Systems.
- Project work: as-is modeling, to-be modeling of core processes sales, communication and implementation of organizational structure and processes, structure of process management, consolidation of the entire process model.

E-Mail: Michael.Laske@t-systems.com

Dr. Redmer Luxem

Senior Consultant, Business Reengineering, Deutsche Bank AG

- MBA (1995) from UGH Siegen and PhD from the University of Münster. (2000). Guest lecturer in Alabama (USA, 1997) and Melbourne (Australia, 2000). Lecturer at the Department of Information Systems, University of Münster (1995-2000). 2000-2001 at the Venture Technology Group, Deutsche Bank AG. Since 06.2002 at Business Reengineering, Deutsche Bank AG. Fields of activity: process optimization and implementation.
- Project work: to-be modeling of asset management, implementation.

E-Mail: prozess@luxem-web.com

Dr. Volker Meise

Bertelsmann DirectGroup

- MBA (1996) and PhD (2000) both from the University of Münster. Lecturer at the Department of Information Systems, Münster (1996-2000). Since 2000 at Bertelsmann DirectGroup.
- Project work: to-be modeling of core process planning and building, development of organizational structure in „real estate management", communication and implementation of organizational structure and processes as well as continuous process management.

E-Mail: V.Meise@bertelsmann.de

Dr. Michael zur Mühlen

Assistant Professor of Information Systems, Stevens Institute of Technology, Hoboken, NJ (USA)

- Master of Business Information Science (1997) and PhD (2002) both from the University of Münster. Lecturer at the Department of Information Systems, Münster (1997-2002). Since 2002 at the Wesley J. Howe School of Technology Management at the Stevens Institute of Technology. Working Group Chair of the Workflow Management Coalition. Director of AIS Special Interest Group on Process Automation and Management. Research areas: workflow management, process- and resource management, process controlling.
- Project work: to-be modeling of core process service, conceptual design of the solution for Intranet-based models.

E-Mail: mzurmuehlen@stevens-tech.edu

Dr. Stefan Neumann

Lecturer, Department of Information Systems, Münster
- Master of Business Information Science (1998) and PhD (2003) both from the University of Münster. Since 1998 Lecturer at the Department of Information Systems, Münster. Research areas: process management, information modeling, trade information systems.
- Project work: continuous process management, process improvement, conceptual support for the SAP implementation.

E-Mail: isstne@wi.uni-muenster.de

Dr. Christian Probst

Lecturer, Department of Information Systems, Münster
- Master of Business Information Science (1997) and PhD (2003) both from the University of Münster (1998). Since 1997 Lecturer at the Department of Information Systems, Münster. Guest lecturer at the Queensland University of Technology, Brisbane (Australia, 2003). Research areas: workflow management, operations management, process- and resource management, service management, trade.
- Project work: interface modeling for parent company of the group and other subsidiaries, modeling of continuous process management.

E-Mail: ischpr@wi.uni-muenster.de

Assoc. Prof. Dr. Michael Rosemann

Director of the Centre for Information Technology Innovation, Queensland University of Technology (QUT), Brisbane, Australia
- MBA (1992) and PhD (1995) both from the University of Münster. Lecturer and Assistant Professor at the Department of Information Systems, Münster (1992-1999). Since 1999 at QUT. Regular guest lecturer at the Business School at Nanyang Technological University Singapore and at the Northern Institute of Technology, Hamburg. Research areas: conceptual modeling, Enterprise Systems, process- and workflow management. Management consultant for process management.
- Project work: modeling conventions, modeling of support processes.

E-Mail: m.rosemann@qut.edu.au

Dipl.-Ing. (FH) Norbert Schnetgöke

Specialist for Process and Management Consulting, T-Systems International GmbH

- Training Telecommunication Tradesman, Study of Electro-Technics with focus on telecommunication technology (1988-94), FH Münster; Fields of activity: IT-development, process modeling and process management, Deutsche Telekom AG (1994-96). DeTe Immobilien (1997-2000). Since 2000 at T-Systems International.
- Project work: as-is modeling and to-be modeling of core- and support Processes, consolidation of the entire process model.

E-Mail: Norbert.Schnetgoeke@t-systems.com

Dr. Ansgar Schwegmann

IT Consultant and Contractor for Prof. Becker GmbH, Münster

- MBA (1994) and PhD (1999) both from the University of Münster. Lecturer at the Department of Information Systems, Münster (1994-1999). Since 2000 self-employed IT Consultant. Fields of activity: design and implementation of IT-systems, organizational re-design incl. business process modeling, process management, process improvement, change management.
- Project work: establishment and maintenance of the modeling guidelines, administration of the ARIS Toolset, provision of Intranet-based models, as-is modeling and to-be modeling in various areas, inter-enterprise process management

E-Mail: info@dr-schwegmann.de

Dr. Mario C. Speck

Manager, CTG Corporate Transformation Group GmbH

- Master of Business Information Science (1996) and PhD (2001) both from the University of Münster. Lecturer at the Department of Information Systems, Münster (1996-2001). Since 2001 Consultant at CTG Corporate Transformation Group GmbH, Berlin. Fields of activity: process management and workflow management, object-oriented information modeling, data management, retail information systems, Internet publishing, call-center simulation, ERP-Software for utility industry, process oriented cost cutting
- Project work: to-be modeling of support processes, modeling of technical terms, consolidation of the entire process model

E-Mail: speck@ctg.de

Dipl.-Kfm. Michael Vieting

Business Consultant, Bayer AG

- MBA (1995) from the GH Essen. Fields of activity: Organization, BPR, SAP, Gesellschaft für Personalwirtschaft und Organisation mbH, Düsseldorf; MAN Gutehoffnungshütte AG, Oberhausen; DeTe Immobilien. Since 1999 at Bayer AG, Leverkusen.
- Project work: to-be modeling of support processes, development of organizational structure for administration and finance as well as personnel and legal affairs.

E-Mail: michael.vieting@web.de

Dipl.-Ing. (FH) Clemens Wernsmann
Head of Process Integration, T-Systems International GmbH

- Study of Mechanical Engineering (1981-85), FH Steinfurt. Fields of activity: Maintenance and Repair, and Process Control, Thyssen Stahl AG, Duisburg (1985-88); System Administrator, Network Designer and Order Management / Workflow, Deutsche Telekom AG, Münster / Bonn (1988-96), Process Management at DeTe Immobilien (1996-2000), Process and Management Consulting at DeTeCSM Darmstadt (2000-2001). Since 2001 at T-Systems and among others cooperator in research projects.
- Project work: project leader of the project "process modeling", design and implementation of continuous process management.

E-Mail: Clemens.Wernsmann@t-systems.com

Deutsche Telekom Immobilien und Service GmbH
Postfach 40 03 · D-48022 Münster
Telephone: +49 251 7770 0 · Telefax +49 251 7770 6999
http://www.DeTeImmobilien.de/
E-Mail: info@DeTeImmobilien.de

Westfälische Wilhelms-Universität Münster
Institut für Wirtschaftsinformatik
Lehrstuhl für Wirtschaftsinformatik und Informationsmanagement
Steinfurter Straße 109 · D-48149 Münster
Telephone +49 251 8338 100 · Telefax +49 251 8338 109
http://www.wi.uni-muenster.de/is/
E-Mail: ls-is@wi.uni-muenster.de

Queensland University of Technoloy
Centre for Information Technology Innovation
126 Margaret Street, Brisbane Qld 4000, Australia
Telephone +61 7 3864 9473 · Telefax +61 7 3864 9390
http://www.citi.qut.edu.au/
E-Mail: citienquiry@qut.edu.au

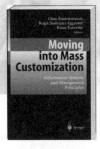